D1421341

Animal Architecture

Also by Karl von Frisch

The Dancing Bees

Man and the Living World

Karl von Frisch | Animal Architecture

With the collaboration of Otto von Frisch

Translated by Lisbeth Gombrich

HUTCHINSON OF LONDON

HUTCHINSON & CO (Publishers) LTD
3 Fitzroy Square, London W1

London Melbourne Sydney Auckland Wellington Johannesburg
Cape Town and agencies throughout the world

First published 1975

Printed in the United States of America
ISBN 0 09 122710 0

Foreword

The idea of this book arose ten years ago when Helen and Kurt Wolff visited us at our country home, Brunnwinkl, on Lake St. Wolfgang, Austria, where for many years I had been collecting specimens of the wildlife in the surrounding countryside. When I showed them my "museum," my publisher friends were particularly interested in some insect nests in my collection, and they suggested that I write a book about animals as architects. Other tasks occupied my time, however, leaving me no leisure for light diversion of this kind. But the idea was by no means forgotten, and now it has taken shape.

I have tried to be generally intelligible, for the book is chiefly intended for the nonspecialized audience. If the public at large knew more about the workings of nature it would help to protect our living environment against the progressive destruction that threatens it.

To understand the building activities of animals it is necessary to know something about the lives of the builders. Therefore the behavior of animals always enters the discussion though only in modest outline so as not to encroach on the space required for the main theme.

I could not have carried out what I had in mind without the help of many friends. B. Hölldobler, M. Lindauer, M. Lüscher, F. Schremmer, H. Sielmann, and others kindly let me have many interesting photos. I have to thank M. Renner for taking further pictures and for other help besides, as well as Mrs. Turid Hölldobler for the faithful illustrations in the text. The Akademie der Wissenschaften und der Literatur in Mainz (German Federal Republic) gave a grant toward the production of the illustrations. It is not possible to mention personally all those who helped, but to all of them I want to extend my warmest thanks, not least to the publishers for the beautiful presentation of the book. Finally I should like to say that the collaboration with my son Otto, who helped me greatly with suggestions and in many other ways, was for me a source of special pleasure.

Brunnwinkl, Fall 1973   KARL VON FRISCH

Contents

Foreword

**THE LIVING BODY AS ARCHITECT—
EXTERIOR AND INTERIOR DESIGNS**   **3**

THE MICROSCOPIC SPHERE   3
SPONGES   9
EDIFICES OF CORAL POLYPS   9
SNAIL SHELLS   14

BUILDERS   **22**

Arthropods   **24**

TRAPPERS   24
    The ant lion   25
    The web of the garden spider   27
    Spiders that hunt in other ways   34
    Underwater nets   41

BUILDING A HOME FOR ONESELF   42
    The larvae of caddis flies   42
    Larval dwellings of Microlepidoptera   45
    The cuckoo spit insect   49

BUILDING TO PROTECT THE OFFSPRING   50
    Digger wasps   51
    True wasps   55
        A solitary true wasp · *Social true wasps* ·
    Solitary bees   66

HOMES OF THE SOCIAL INSECTS   72
    Bumblebees' nests   73
    Honeybees and their dwellings   81
        *Homes that beekeepers provide and bees fit
        out* · *The bee community* · *Construction of
        the comb* · *Measuring instruments of the
        bees* · *Orientation of combs by the earth's
        magnetic field* · *Bees' glue (propolis)* · *Mov-
        ing house* ·

Homes of the ants   96
Castes of the ants and their tasks · From modest soil dwellings to stately mounds · Dwellings made of wood or paper · Ants as weavers · Storage chambers and culture rooms · Building roads and animals sheds · Vagrants without fixed abodes ·

Termites, masters in building and civil
    engineering   123
The termite community · Simple structures · The great architects · Air-conditioning in termite dwellings · Building techniques ·

Vertebrates   **151**

FISHES   152
Salmon as modest builders · Bubble nests of labyrinth fishes · The refuge of the sand goby · Nests of sticklebacks and wrasses · Jaw-fishes · Mouthbrooders ·

AMPHIBIA   165
Foam nests · A tree frog builds with clay · A strange nursery for little frogs ·

REPTILES   169

BIRDS   171
Birds that build and regulate incubators   172
Breeding without nest-building   183
Simple nests   185
General remarks on nests and nest-building in
    birds   188
Cup-nests   191
A roof over one's head   201
A hanging nest protects against unwelcome guests · Weaverbirds ·
Communal nests   211
Birds as tenants   217
Hole-breeders   218
Woodpeckers · Hornbills · Kingfishers · Birds that build clay shelters ·
The tailorbird   227
Edible nests of swiftlets   227
A living nest with central heating   231
Bowerbirds and their bowers   237
What passes in the mind of a bowerbird when he
    builds and decorates his bower?   244

MAMMALS   247
   The mole and its subterranean environment   248
   The badger   251
   Rodents as builders   253
    The harvest mouse · The dormouse · Wood
    rats · Squirrels · Marmots · Beavers ·
  Apes   278

Conclusion   **283**

Table of Metric Equivalents   **289**

Picture Sources   **291**

Subject Index   **295**

Index of Scientific Terms   **305**

Animal Architecture

When we stand before great churches, temples, pyramids, and other works of architecture built hundreds, if not thousands, of years ago, our minds are filled with awe and admiration. Yet there have been architects millions of years before that. Their work, it is true, owes its existence not to the inspired genius of great artists, but to the unconscious, unremitting activity of the force of life itself. Without tools, indeed without anything that could be called action, the coral polyps of the warm seas erected their limestone piles—edifices that can reach the size of mighty mountains—and they go on building today. Certain microscopic organisms, the Radiolaria, have been producing glasslike supporting structures for their tiny delicate bodies for even longer periods. Living dispersed as they do over the vastness of the oceans, they do not build up imposing monuments out of their siliceous skeletons; but many an artist's eye has been transported by the contemplation of their exquisite beauty. To such organisms we shall briefly turn our attention.

But mainly this book will be devoted to the activities of animals that actually build structures of the greatest diversity from extraneous materials or from substances they produce within their bodies—using techniques akin to those that humans employ in masonry, weaving, plaiting, digging, and so on. Some of these structures serve as traps for prey, but most of them are intended as protection for the animal's own body or for its young. Nature has provided these builders with the tools of their trade: they use their teeth, their beaks, their legs, and other parts of their bodies. In many cases these organs are amazingly well adapted to the special tasks they have to perform.

THE LIVING BODY AS ARCHITECT—EXTERIOR AND INTERIOR DESIGNS

THE MICROSCOPIC SPHERE

At the bottom of the animal kingdom we place the unicellular animalcules, the protozoa. As a rule they are so
small that they are invisible to the naked eye, or nearly
so. One of the most primitive among them is the amoeba,
an inhabitant of fresh-water puddles. Its body consists of
a little blob of protoplasm and a nucleus. However, even
the organization of amoebae is not quite so simple as
previously had been assumed. The electron microscope
has revealed that many elements of structure in the protoplasm and the nuclei of unicellular animals are not very
different from corresponding formations in the cells of
higher animals. In comparison with these, of course, the
organization of amoebae is very primitive. The amoeba
can move along the leaf surface of an aquatic plant—or
any other surface—in any direction by allowing its protoplasm to flow in that direction, extending so-called pseudopodia (Greek for "apparent legs") and retracting
similar pseudopodia from another direction (fig. 1). An

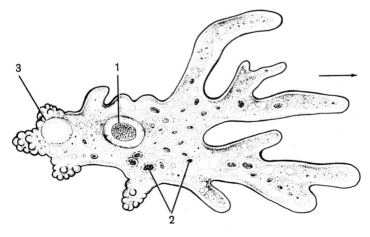

Fig. 1. Amoeba (Amoeba proteus)
crawling in direction of arrow.
(1) Cell nucleus; (2) food residues;
(3) contractile vacuole whose
rhythmic contractions pump out
the water which constantly perme-
ates the organism. Size in the
longitudinal direction, ½ mm. In
good light, amoeba may be visible
to the naked eye as a minute white
speck.

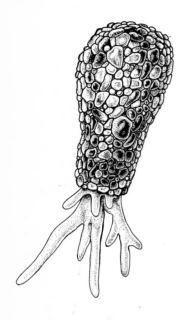

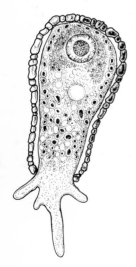

amoeba is capable of letting its protoplasm flow around small particles of food, for instance tiny algae, and incorporating them. For this it needs no permanent mouth or orifice. Multiplication takes place when an amoeba has grown to twice its original size. After the nucleus has divided, the protoplasm splits into two parts of equal size. The "mother" has turned into two "children." Another protozoan, *Difflugia,* incorporates not only food but such indigestible particles as grains of sand. These particles migrate from the interior of the body protoplasm to the exterior where they are cemented together with a substance secreted by the animal. The *Difflugia* eventually comes to sit within a protective urn-shaped casing, or "test." With the aid of pseudopodia protruding from the opening of this test, it can crawl about like a snail with a "house" on its back (fig. 2). In this species, as in the amoeba, propagation is by division of nucleus and protoplasm; division never takes place, however, before a second little house is ready. To begin with, some of the protoplasm flows out of the opening, takes on the urn shape of the casing, and remains immobile until the basis of a new casing has formed on its surface. Only then does the animal divide into two daughter organisms—each crawling away with its own miniature home.

The ingestion of indigestible particles in addition to edible ones and their use for the building of a "house" in this roundabout way is highly original and unique in the animal world. In another group closely related to *Amoeba* and *Difflugia,* the large and diverse marine order of Foraminifera, the method is different. The Foraminifera build their casings out of lime. The protoplasm of these tiny builders is able to extract from sea water the low concentrations of carbonate of lime ($CaCO_3$) dissolved therein (about 0.35%), concentrate them further, and precipitate them in a special form as calcite (limestone). Most species grow to many times their original size before they divide. They start with a single chamber, but later add many more (fig. 3). The dividing walls, or septa, between these chambers, and frequently also their outer walls, are perforated. Hence the name of "Foraminifera" (from Latin *foramen,* "hole," and *ferre,* "to carry or bear"). The protoplasm in these chambers is connected by way of apertures in the separating walls and sends forth delicate, finely branched pseudopodia in search of food. Multiplication is prepared some time in advance by a series of repeated divisions of the cell nuclei; finally the protoplasm,

Fig. 2 Difflugia pyriformis. *Below, a longitudinal section. This amoeba builds around itself a casing, or test, from grains of sand.*

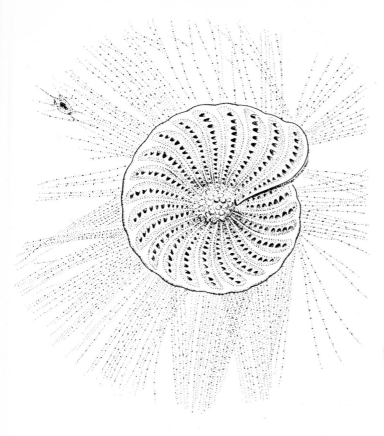

too, breaks into many small parts that leave the old shell as uninuclear daughter individuals, which start building their own little homes. On sandy shores, where Foraminifera are plentiful, one will often be surprised to find that many particles one has taken for grains of sand turn out under magnification to be foraminiferous shells. As many as fifty thousand have been counted in one gram of sand. In some parts of the world, their remains have accumulated so much in the course of the earth's history that they form part of certain geological strata.

No magnifying glass is needed to recognize the fossils derived from the genus of Foraminifera that has given its name to nummulitic limestone, for some of them are up to six centimeters in diameter. These giant unicellular organisms lived in the Tertiary period, some fifty million years ago (pl. 1, p. 6).

The morphological variety of foraminiferous tests is

Fig. 3. Foraminifera of the genus Polystomella. *Left, a living animal- cule. Protoplasm fills the interior, covers the outer casing also, and sends forth thin filaments, or pseudopodia. In drawing above, these have caught a minute organ- ism which is being digested by the protoplasm. Right, a diagrammatic cross section through the calcareous test. The partitions, or septa, between the chambers, and the outer casing are perforated. Diameter about 1 mm.*

Plate 1. Nummulitic limestone. Many small nummulites (Assilina exponens). *Top right, a nummulite* (Nummulites millecaput) *having a diameter of 5½ cm. See figure 4:5 for structure. Found at Höllgraben near Adelholzen, Upper Bavaria.*

illustrated in figure 4. Their shapes differ from species to species, suggesting the existence of similar specific differences in the protoplasms from which they originate. The nature of the inner forces responsible for the specific differences in the forms of the skeletons, however, is one of the many unsolved mysteries in the realm of life.

In other groups, the differences between species may be even more striking. This is the case in the Radiolaria, the nearest relatives of the Foraminifera. They form part of the floating plankton of warm seas, and the substance they extract from sea water is not lime but silica, which occurs in an even more diluted concentration than lime. From this substance, they fashion in the interior of their bodies protective and supporting structures of marvelous beauty. (Many years ago, a great nineteenth-century biologist, Ernst Haeckel, described the abundance and variety of their exquisite forms in two works: *Radiolaria,* 1862 and 1887, and *Kunstformen der Natur* ["Nature's Art Forms"], Leipzig, 1899.) Chemically, these siliceous skeletons resemble glass, and might be considered, therefore, a more delicate substance than the calcareous skeletons of the Foraminifera. Their shapes may truly be called noble. Though in some species the skeletons consist merely of a loose agglomeration of stipules, in most species they form elaborate structures similar to little helmets, or to latticed balls, which are often intricately interlaced, and to many other configurations (fig. 5 and pl. 4, p. 17). I do not want to wax philosophical about so much "useless" beauty scattered over the oceans—Nature is prodigal: she almost matches the beautiful intricacies of these tiny animate forms with the exquisite shapes of snow crystals in the inanimate world.

When a cell divides, loose skeletal stipules can be transferred to daughter animals. But latticed balls and other rigid structures cannot be divided. Often, one of the two daughter animals inherits the whole skeleton while the other gets nothing and has to start building from scratch. Sometimes the animal breaks up into many tiny daughter organisms, as with the Foraminifera. When this happens, each daughter builds itself a new home, while the abandoned skeleton sinks slowly to the bottom. In view of the prodigious age of this group (they date back to the Precambrian era—about seven hundred million years ago) and the durability of their structures, it is hardly surprising that in some parts of the world the ooze at the bottom of tropical seas consists predominantly of their skeletons.

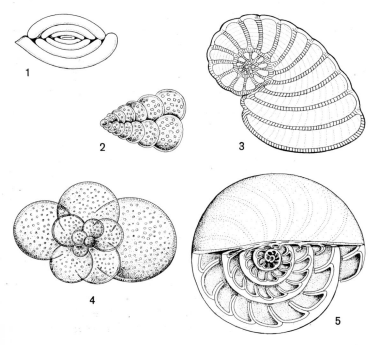

Fig. 4. Some examples illustrating the variety of forms of foraminiferous tests: 1, Miliola; 2, Textularia; 3, Peneroplis; 4, Globigerina; 5, Nummulites. The test of Nummulites has been partly cut open to reveal the interior chambers.

7

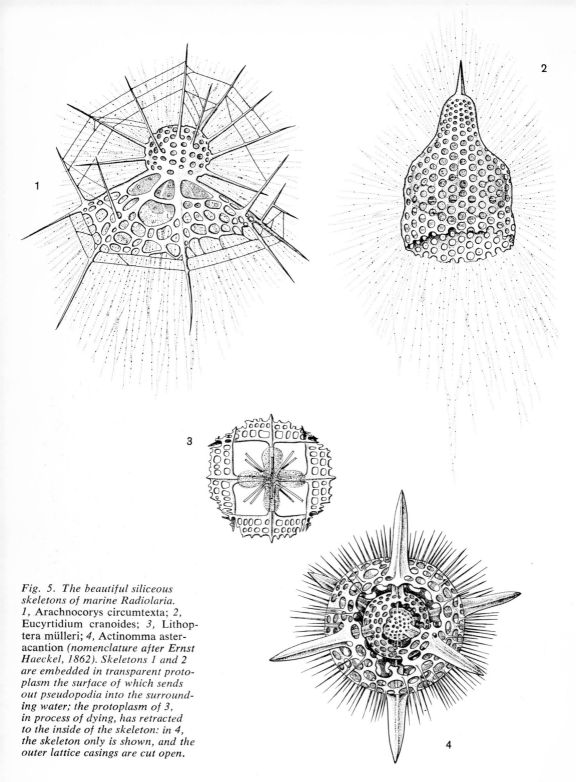

Fig. 5. The beautiful siliceous skeletons of marine Radiolaria. 1, Arachnocorys circumtexta; *2,* Eucyrtidium cranoides; *3,* Lithoptera mülleri; *4,* Actinomma aster- acantion *(nomenclature after Ernst Haeckel, 1862). Skeletons 1 and 2 are embedded in transparent proto- plasm the surface of which sends out pseudopodia into the surroun- ding water; the protoplasm of 3, in process of dying, has retracted to the inside of the skeleton: in 4, the skeleton only is shown, and the outer lattice casings are cut open.*

SPONGES

To proceed from the minute and delicate protozoa to the sponges may seem capricious. For sponges of the kind sold in drugstores look very different from Radiolaria. They are of considerable size and once formed the supporting skeleton of animals which, when alive, anchored themselves firmly to submerged rocks in the sea. Sponge divers collect them from depths of thirty meters or more. A fresh sponge is slimy to the touch. Its soft body consists of innumerable cells penetrating and enveloping the elastic "spongy" skeleton made of a keratose substance called "spongine." The name reflects the origin of the substance but tells us nothing of its chemical nature: in fact, it consists of proteins. Other genera and species of sponges construct their skeletons from calcareous or siliceous substances. We note similarities to Foraminifera and Radiolaria, especially in that their calcareous or siliceous stipules are also formed inside the cells initially. Soon, however, these stipules grow bigger than the cells in which they originated. Other cells attach themselves and secrete more material. In this way, ray- or fan-shaped groups of stipules, and a great variety of other forms—balls, crosses, anchors, etc.—may arise. In some sponges the skeletal stipules are cemented together to form a rigid structure as, say, in the deep-sea sponge *Euplectella aspergillum* (pl. 3, p. 17). From its base, firmly anchored to the bottom, it may reach a height of half a meter. In the turbulent waters near shore, its vitreous chalices would quickly break, and that is why these animals are only found in deep waters.

EDIFICES OF CORAL POLYPS

Polyps are slightly more differentiated than sponges, but the general plan of their bodies is still very simple. The fresh-water polyp (*Hydra*) consists essentially of a delicate tube a few millimeters long, attached at its closed end to some surface such as the leaf of an aquatic plant, and bearing at its free end a set of tentacles around a central orifice that serves as both mouth and anus (fig. 6). The whole inside of the tube is a stomach. The tentacles are equipped with microscopic poisonous weapons (stinging cells) enabling the *Hydra* to kill and devour quite large morsels of food, such as the tiny crustaceans *Daphnia* and *Cyclops*. The orifice can be extended enormously.

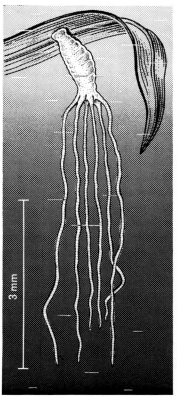

Fig. 6. A fresh-water polyp (Hydra).

3 mm

The sea harbors many species of larger polyps. The sea anemones (*Actinia*) belong to the class of Anthozoa, or flower animals. And, indeed, these creatures, sitting on the sea floor with their colorful bodies and their crowns of tentacles spread like petals, resemble flowers rather than voracious animals. Nothing is hard in the body of a sea anemone. But other species of Anthozoa, the coral polyps, form calcareous skeletons.

In figure 7 (top right) this process is shown on a single animal. It secretes a calcareous substance from its bottom surface, the skin of which exhibits many pocketlike folds, the result being a delicate ribbed pattern. The skeleton differs from that of protozoa and sponges in that it is an *exterior* skeleton. The polyp provides itself with a calcareous pediment which gradually lifts it higher and higher. The skeletons of coral polyps living isolated lives —and only a few species do so—consist entirely of this pediment which increases in width from the bottom up because the polyp grows both in size and width in the course of its life. Animals of the genus *Fungia* may reach a diameter of twenty-five centimeters; they are giant polyps sitting on imposing pediments.

In other species the growing polyp divides into two individuals, each of which proceeds to form its own skeleton. Successive divisions lead to the characteristic arbuscular coral structure shown in plate 7 (p. 18). In the upper part of this picture is a polyp in process of division; two have just divided and formed skeletons: the others are already branching away from each other. Though most species of coral polyps form much larger colonies with tightly packed individuals (a detail of this type of colony is shown in fig. 7), the principle is the same. The body walls and gut cavities of neighboring polyps remain connected. Hence the skeleton follows the branching of the soft bodies that form it. In the course of upward growth, the polyps pull up their lower end from time to time, and then proceed to secrete a new calcareous disk, forming a transverse layer. In this way, corals may grow several centimeters in a year and eventually develop into huge blocks with thousands, even millions, of individuals. The individual animals remain small. They usually measure one-half to one centimeter in diameter. But they grow and multiply incessantly, and a colony may reach an age of many decades. In addition to progagation through division, coral polyps also reproduce sexually. This leads to the release of swarms of larvae and the

formation of new colonies. Their shapes may vary considerably because the manner of dividing, or "budding," differs between species. Some coral structures are tightly branched, others are loosely branched; some form round blocks, others form shapes resembling disks, or candelabra. Plate 11 (p. 19) gives an idea of this variety. It shows a small part of a coral reef as it emerges at low tide. The polyps have retracted completely and have covered their skeletons, as if with delicate skins. By secreting a mass of mucous substance, they protect themselves from dehydration and can withstand several hours' exposure to the air without suffering any injury. As soon as the water returns, they will stretch themselves once more and spread their tentacles. A sea of blossoms seems to have opened in plate 8 (p. 18). The scene becomes truly magical through the rich and colorful population of fishes and other animals for whom the fantastically shaped coral structures provide hiding places and protection against innumerable predators. The coral polyps themselves are well armed against enemies through their stinging cells; moreover, in the face of danger they can withdraw with lightning speed into their limestone skeletons. But they, too, have enemies. Arms races are not a human invention. Everywhere better methods of defense lead only to better aggressive weapons. Thus parrot fishes

Fig. 7. A small section of a colony of coral polyps. One polyp expanded; two partly expanded; one wholly contracted; below, right, a skeleton after removal of soft tissue. Diameter of the pediment is about 8 mm. Top right, diagrammatic cross section of single polyp slightly magnified. The ribbed foot secretes lime on its underside, thereby forming pediment on which the polyp rests.

have developed strong teethlike beaks which enable them to bite off and crunch whole branches of corals in order to feed on the polyps; nor are they the only ones capable of enjoying this stony pasture.

Since about 1963, coral reefs have faced a serious danger from a new enemy. A starfish (*Acanthaster planci*), which because of the arrangement of its dorsal spines has been given the name of "Crown of Thorns," is spreading on the reefs in large numbers. It systematically eats the coral polyps, leaving the dead skeletons behind. Should the starfish population go on increasing at this rapid rate, the result could be the destruction of corals over wide areas, and the consequent danger to many coasts for which coral reefs provide protection from heavy seas.

Reef-forming species live only in warm seas; they require a water temperature of at least 20° C. (68° F.). Conditions are particularly favorable for them near the coasts and islands of the Pacific and Indian oceans. Fringing or shore reefs, formed by coral polyps, may follow the surf line of a coast for hundreds of kilometers, or skirt it from a greater distance—up to 150 kilometers (pl. 9, p. 19). They may surround islands in mid-ocean, or form circular atolls where there are no islands (pl. 10, p. 19). They sometimes reach up from depths of four thousand to six thousand meters. This phenomenon was once most puzzling to scientists because coral polyps were known to thrive only to depths of forty to fifty meters. Coral polyps need light to nourish certain plant organisms, unicellular algae, that live inside their cells—an association (symbiosis) of benefit to both partners. How, then, could coral limestone be formed at such great depths?

The explanation suggested by Charles Darwin is still largely accepted today and has been supplemented or modified only in details. Coral limestone is indeed formed only close to the surface of the sea, the sphere in which the polyps live. Where the calcareous masses reach down to greater depths, the bottom of the sea had subsided slowly—as it still does in many parts of the world. To the extent that the ground sinks, a new life zone for corals is added at the top, and this is soon filled with new coral colonies. In this way, contact with the surface is maintained, while lower down the colonies die and only their limestone structures remain as building blocks of the reef. From the surf region downward, the gaps in these structures are filled progressively with broken coral, snail or bivalve shells and such, which other calcium-secreting organisms, especially algae, busily cement together.

The polyps had plenty of time for their building program. Borings at great depths near the Bahamas have indicated that the coral reefs of that region have been growing since the middle Cretaceous era or longer, that is to say, at least eighty million years. This process of growth over long periods of time, coupled with the slow subsidence of the sea bottom, presents an explanation for all the different forms of reef. Fringing reefs form where rocky ground in shallow depths close to the shore affords coral polyps a base on which to settle and grow upward. If the sea bottom sinks and the depth of water increases, the reef may continue to grow, but because the coast will have been partially submerged, the distance between reef and coast will have increased, thus forming the barrier reef (p. 9, p. 19). And when an island surrounded by a coral reef gradually sinks into the sea, the remaining formation is the atoll (fig. 8).

Conversely, the bottom of the sea can rise as well as sink. Old sea bottom can become dry land and can even be folded into mountain ranges. This is why, when keeping one's eyes open on mountain rambles, it is often possible to find fossil marine snails and other shells at very high altitudes. Fossil corals also turn up. In the Austrian Alps, the Bischofsmütze ("Bishop's Mitre Mountain") in the Donnerkogel range, which forms part of the Dachstein massif, is one such old coral reef. This great limestone range was built up by coral polyps for the most part, eons ago when there was sea all around and temperatures were very different from those of today.

There is no need to travel to the tropics in order to observe coral animals. True, the species inhabiting temperate zones do not produce such great and indestructible edifices as their tropical relatives do, but some of them have interested man, nevertheless. Travelers to Italy may recall colorful necklaces and other pieces of jewelry made of red coral being sold at stalls in Venice, and elsewhere. This jewelry is fashioned from skeletons of the red precious coral *(Corallium rubrum)*, a species frequent in the Mediterranean. Their skeletons, which are only moderately ramified, grow to a height of twenty to forty centimeters. Few purchasers of coral ornaments know what coral polyps look like when alive and growing on the bottom of the sea. Like the coral animals of the reefs, these polyps extend their bodies and spread their ten-

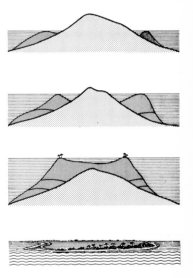

Fig. 8. The development of an atoll. Top, an island with a fringing reef. As the bottom of the sea gradually subsides, corals grow upward. Those covered by more than 50 m. of water perish. Bottom: if the process of subsidence continues, the island becomes submerged and only the ring-shaped coral atoll remains.

tacles in search of food, but at the slightest disturbance they withdraw with lightning speed into their hard skeletons for protection and support (pls. 5 and 6, p. 18).

SNAIL SHELLS

We are justified in calling a snail's shell its house: it is a mobile home that protects a large part of the animal's soft body and usually allows it to withdraw completely at the first sign of danger.

Though in most species the mature shell is coiled spirally (fig. 9), it does not start that way. During the animal's development, the shell appears first as a little calcareous cap on its back. For in contrast to coral polyps, snails form their calcareous structures on their backs. However, in both coral polyps and snails the production of the hard substance is concentrated in a special part of the skin. The necessary raw material, carbonate of lime, is available to the snail in dissolved form in its food and in water; it is subsequently secreted by the skin in the form of aragonite or calcite.

Snails are much more differentiated than polyps and possess clearly distinct body parts. The head, in front, bears one or two pairs of feelers (pl. 13, p. 20), which serve the chemical and tactile senses, and, in the Roman, or the edible, snail, have even an optical function, for the club-shaped ends of their upper pair of feelers carry a pair of very primitive eyes. The base of the body, its "foot," is muscular (fig. 9); its undulating contractions propel the animal along any surface. Of particular interest are the changes that take place in the shell in the course of the animal's development. As they grow, the viscera, consisting of intestines, digestive glands, heart,

Fig. 9. Roman, or edible, snail (Helix pomatia). *The edge of the mantle with the breathing hole is visible below the rim of the shell. The two upper (longer) feelers carry simple eyes on their tips.*

kidneys, and sexual organs, bulge upward and backward in a manner suggestive of a hernia. It is the skin covering this developing pouch, the so-called mantle, that secretes the shell. Since the pouch tends to curl spirally in the course of its development, the shell too adopts this form. A fold grows down from the mantle enclosing (in the air-breathing species) an air-filled space, the snail's lungs, connected with the air outside by means of a breathing hole. This can be seen in figure 9 just below the lip of the shell, right in the middle of the thickened edge of the mantle, which protrudes from under the shell. The opening is the only place where the shell grows and where the edge of the mantle periodically adds new layers of calcareous substance. The individual growth layers can be clearly seen. Away from this zone of growth, the shell is only slightly thickened by further calcareous secretion from the inside.

The class of Gastropoda to which the snails belong contains over a hundred thousand different species found on land, in running or stagnant water, and in particular abundance in the sea. Not all of them possess shells, and in some their shells are not visible from the outside and are no more than rudiments embedded inside the soft parts of the body. Hence what I have described so far does not apply generally; but in our context we are only interested in "house-owners." In the garden, or the edible, snails and many other species the thickened border of the mantle and hence the lip of the shell opening is simple and smooth. But anyone picking up snail shells thrown up onto the beach by the sea will encounter a much greater variety of forms. It is for good reason that snails have attracted as many collectors as have postage stamps, coins, or beetles. Their whorls may be coiled in various ways; they may be flat like a disk, slightly raised in the shape of a flat cone, or piled high like a steeple; their interior space may increase gradually or rapidly; they may sport various ornamentations, such as scallops or spikes (pl. 12, p. 20). All these variations originate during growth at the rim of the shell in response to corresponding configurations of the mantle border, which alone determines form. This applies also to the snail on the right-hand side of the upper row in plate 12. Its strange-looking excrescence is secreted by a forward-pointing tubular appendix of the mantle border. The "siphon" so formed serves to conduct fresh water to the snail's breathing organs (in these aquatic snails, their gills). The deposition of pigments, which often provide snail shells with lively patterns, also takes place in the mantle border. These colors may delight collectors, but because of their poor vision the snails themselves are hardly in a position to appreciate their own beauty.

Just a quick glance in passing—at a tangent, as it were—at the multifarious world of worms. Among the more highly developed bristle worms (Polychaeta), there are certain species that fashion protective tubes around their bodies from lime or from cemented sand particles. In these tubes they spend all their lives, firmly attached to the ground or to an aquatic plant. The primitive structure of the tubes is devoid of any architectural ornamentation, but this only serves to enhance the charm of the delicate crown of tentacles protruding from the animals' heads in the opening of the tube into the water above (pl. 2, p. 17, *Megalomma*). The whirling movements of these threadlike extensions, bearing microscopic cilia, propel minute particles of food into the central mouth opening. The black dots near the end of the threads are primitive eyes. In the event of danger, the whole crown of tentacles suddenly disappears into the tube.

A garden snail can live for six to seven years. But it is not equipped for life in winter. In the cold season the animal withdraws completely into its shell and seals its opening by secreting a calcareous plug which is expelled only when spring returns. In the meantime, the animal hibernates, fast asleep. Some terrestrial and aquatic snails have developed a more elegant method of closure, one, moreover, that is available at all times as a protection against predators. They possess a round disk, or operculum, on the back of their tail region which at first sight appears to be quite unrelated to the shell (fig. 10, left).

Fig. 10. The marine snail (Murex brandaris). *Close to the tail end is a round disk formed by skin secretion, which exactly fits into the opening of the shell when the snail withdraws into it.*

Plate 2 (right). The tube worm **Megalomma** *constructs a protective tube from cemented grains of sand. The expanded crown of tentacles protrudes from the opening in search of food. (See p. 16.)*

Plate 3 (above). The vitreous skeleton of Euplectella aspergillum *("Venus's flower basket"). It is formed inside the body of the sponge from silicates extracted from sea water. Length 24 cm. Other specimens may reach up to 60 cm. in length. (See p. 9.)*

Plate 4. Microscopic siliceous skeletons, the remains of the unicellular Radiolaria, form vast accumulations at the bottom of warm seas. The colors of the photograph are the effect of special illumination. (See p. 7.)

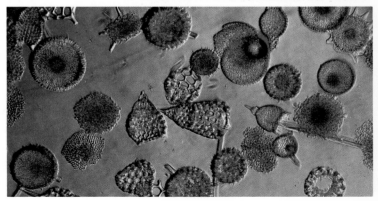

17

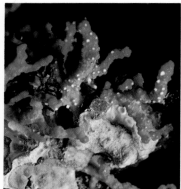

Plate 5. *Red precious coral* (Coral-lium rubrum) *showing polyps retracted.*

Plate 6 (right). *The same coral with polyps extended. Mediter-ranean. (See p. 14.)*

Plate 7. *Skeletal structure of a small coral colony showing stages of division. (See p. 10.)*

Plate 8 (right). *Coral polyps under water, expanded. Part of a colony of* Goniopora. *Shadwan Island, Red Sea. (See p. 11.)*

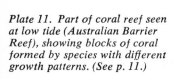

Plate 9. Barrier reef in the South Pacific (Society Islands). Ships can enter through the gap into the calm waters of the lagoon. (See pp. 12, 13.)

Plate 10. Atoll. Viti Levu Island, Pacific Ocean. Aerial photograph. (See p. 12.)

Plate 11. Part of coral reef seen at low tide (Australian Barrier Reef), showing blocks of coral formed by species with different growth patterns. (See p. 11.)

19

Plate 12. A selection of snail shells to give some idea of their variety. Upper row, left to right: Cyclophorus eximius, *Javanese terrestrial snail; (all following species are aquatic)* Scala scalaris *(Pacific, China);* Charonia tritonis *(Indo-Pacific);* Tibia fusus *(near China). Lower row:* Astraea longispinum *(West Indies);* Murex palma-rosae *(Indian Ocean);* Lambis scorpius *(Indonesia, western Pacific);* Murex tenuispina *(Indo-Pacific). (See p. 15.)*

Plate 13. A fresh-water snail. Its shell is colonized by pedunculate unicellular organisms. The colony is one of bell animalcules (Carchesium). (See p. 14.)

But we only need to frighten the animal a little to make it withdraw into its shell: we will then notice, to our surprise, that the disk is positioned in such a way that it reaches the shell last, sealing the opening perfectly (fig. 10, right). This lid consists of lime, as does the shell itself. In some species it is made of a keratose substance. Its concentric rings show that as the shell grows the lid grows too, so that it always fits.

Snail shells are often beautiful; and being hard, they provide excellent protection. Their beauty has aroused the interest of collectors; their hardness has attracted an interested party of an entirely different kind. Life in the sea is so abundant that most dwellers at the sea bottom suffer from a chronic housing shortage. Hermit crabs have found an original solution to the problem: they search for an empty snail shell of suitable size, and having found one, move into this ready-made accommodation. As a result of so doing through untold generations, they have become so adapted to their dwelling that their tails have taken on the spiral form of a snail's shell, and the familiar hard armor of the crustaceans' rear body that we recognize in ordinary crabs or lobsters has given way to soft skin. Only the animal's head and legs and its formidable claws protrude from the opening. These claws provide as effective a closure to the mouth of the shell as the lid of its former rightful owner.

From the interior skeletons of protozoa and sponges, the calcareous pediments of corals, and the shells of snails we could follow a long continuous trail, looking systematically at the protective and supporting skeletal structures of other animals and finally at our own human skeleton. Such is not my intention. But I thought it appropriate briefly to draw the reader's attention to the beautiful and well-adapted structures found in the lower forms of animal life and present also in the most highly developed organisms—structures that needed no conscious effort from their makers and that existed long before the first master builders and artists consciously set about to create useful and beautiful things.

BUILDERS

Men nowadays tend more and more to leave the work of building and construction to the machines they have invented for the purpose. But in former times a man's handiwork was really the work of a man's hands. The tools he used were simple, but often perfectly shaped for their purpose. Plowshares and hammers are the products of accumulated experience handed down from generation to generation over untold years, and the origin of tools goes back to the very roots of our race. *Australopithecus,* who lived about one million years ago and who is considered the most primitive representative of the family of "man" (Hominidae), did not yet know the use of fire, but he knew how to fashion pebble tools from stone and how to provide them with sharp edges by chipping. This was the first known attempt to produce a kind of universal tool that could be used for striking, scratching, scraping, and cutting.

The development of tools to ever higher degrees of perfection has been the prerogative of man. Among animals, the use of tools that are not parts of their bodies is rare. They mostly use the organs of their bodies, chiefly their mouth parts and their legs.

The most usual purpose of building activities in animals is to make a home that will give protection. Such a home may be constructed for the building animal itself, for its progeny, for the family as a whole, or, by social cooperation, for large colonies as, for instance, in the case of the social insects. The enormous morphological differentiation of animals and the great differences in their needs and faculties are reflected in the great variety of the homes they build. But homes are not the only things they construct. Some animals are trappers that dig pitfalls or weave nets. The roads built by ants and termites, and the remarkable dams erected by beavers to regulate the water of streams to suit their needs, are well known. A description of all structures made by animals in all their various forms would clearly be an endless task. I have had to select, and I have restricted my choice by taking my examples from two groups of animals only. These will be treated separately.

It goes without saying that we shall look at the constructions of vertebrates, for these are closest to ourselves—man belongs to the vertebrate class of mammals. Everybody has seen the nests of birds and would, I expect, like to know more about them. However, a great many people may be surprised to learn that some of the lower vertebrates, that is, certain reptiles, amphibians, and fishes, also build structures of a kind.

The other great group which we shall consider is that of the arthropods, and, more particularly, the insects. They deserve to be included, for their anatomical complexity and the performance of their organs equal that of the higher vertebrates. Insects are by no means "low forms of life," though this is a description often given to flies and wasps. Their phylogenetic evolution proceeded differently from that of the animals with backbones, and for this reason their skeletons, their sense organs, their breathing apparatus, etc. have come to solve their respective problems in a manner that is different from, but in no way inferior to, that of the vertebrate group. In certain respects their performances are even superior to our own. It will, therefore, come as no surprise to learn that some of their structures are highly original, that they frequently differ in their materials and manner of construction from those erected by vertebrates, and that they may reach very high levels of perfection.

When human beings start to build, they first make a plan and try to find the best solution for each individual case. Animals do not need all that. They follow innate drives. Even the greatest architects among them work correctly by instinct. This applies not only to arthropods but, in general at least, to the vertebrates as well, notwithstanding the fact that the course of phylogenetic evolution of the central nervous systems of these major groups proceeded by entirely different pathways, and that only in vertebrates did it lead to that predominance of brain development which, finally, in man, gave rise to actions based on reason and insight. But traces of such higher mental activities seem to be apparent in the building work of some birds and mammals.

Arthropods

In their endeavors to collect and to classify, zoologists have so far described something like a million and a half different species of animals and sorted them into phyla and their hierarchic subdivisions—classes, orders, families, genera, etc.—by reference to similarities and presumed evolutionary relationships. The phylum Arthropoda ("Jointed-legs") derives its name from the highly developed articulation of the legs of its members. Of its classes, the three comprising the greatest number of species are well known to the layman: they are Crustacea (shellfish and their relations), Arachnida (spiders, ticks, and mites), and Insecta (insects), the largest class by far. About one million different species of insects have been described so far. Insects therefore represent approximately two-thirds of all known animal species.

TRAPPERS

Fig. 11. The ant lion (Myrmeleon formicarius). *Left, the larva, seen from the underside; right, the fully developed insect.*

Trapping as a means of catching prey is not a very common method of obtaining food in the animal world. The two examples that follow are designed to illustrate that some trappers achieve their purpose with very simple devices, and others with amazingly elaborate contraptions.

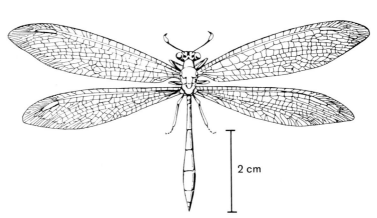

4 mm

2 cm

The ant lion

Ant lions are insect larvae just like caterpillars or maggots that hatch from butterflies' or horseflies' eggs and turn into winged adults after pupation and metamorphosis. They belong to the genus *Myrmeleon* in the order of Neuroptera, and as adults they look rather like damselflies (fig. 11, right). However, their mode of life has little in common with that of damselflies and their beautiful relations, the dragonflies. They do not lay their eggs in water, but search for dry sunny spots, say under the exposed roots of some roadside tree where their larvae can construct their funnel-shaped pit traps in fine sand or dust. In favorable places, great numbers of these pits can often be found side by side. At the bottom of the funnels the predaceous larvae are well dug in—only their flat heads with their wide-open, pincerlike jaws are visible (fig. 12).

In warm weather there will be ants scuttling about in such places. If one of them should happen to get too close to the edge of a funnel, it will slip down. Sometimes it will land straight in the open pincers which will snap to with lightning speed. Sometimes the ant will try to beat a hasty retreat, but before it can once more gain the edge of the pit, the ant lion, making quick jerking movements with its head, showers it with repeated salvos of sand. They may either hit the fleeing ant itself, or they may cause the loose sandy slopes of the funnel to slip, the landslide taking the insect irresistibly with it to its doom. Escape is rare indeed.

When the voracious lion digs its pointed jaws into its victim, it injects a deadly poison. But it does not proceed to dismember its prey, for its mouth parts are in no way adapted to such a task: it has its own highly efficient and elegant method of feeding. Through a groove running inside its mandibles it injects into the body of its prey its gastric juices, juices which normally do their work in an animal's stomach or gut on food that has been ingested. As the muscles and intestines of an ant are contained within a strong chitinous exoskeleton, the soft parts of the victim become completely dissolved in the course of a few hours without a drop being lost. The dissolved nutrients are then sucked up through the same groove in the ant lion's mandibles; the ant's empty, indigestible carcass is hurled out of the funnel. A few more energetic jerking movements with the head get rid of any

Fig. 12. The ant lion lying in wait in its pit. Below, diagrammatic sketch.

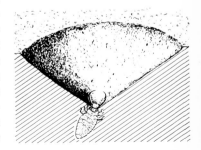

surplus sand accumulated at the bottom, and the funnel is once more ready for the next hapless arrival. Though ants do represent the major part of the ant lion's prey, occasionally its diet is varied by the inclusion of some other straying insect or small spider.

How long the larva will take to grow up, to pupate after a series of moltings, and to conclude its rapacious career will depend on its success as a predator. If it is not well nourished, the larval phase may last two years. However, no matter how long it stays in its funnel, it never fouls the sand of its dwelling with feces for the simple reason that its bowels are closed at the end. By digesting its food outside its body, the larva has no residual waste substances to get rid of.

In former times, the ant lion was looked upon as a very cunning animal. Its actions seemed eminently practical and purposeful. In fact, they are based on a few innate drives and a small number of stereotyped reflexes constantly repeated. These the animal carries out to perfection without ever having been shown how, and without the need to learn from experience. If one snatches one of these little animals from its funnel by a quick movement and places it in a container filled with dry sand, one can study its behavior at leisure.

At first, it will play dead for a few minutes. Since its color is that of sand, keeping still is its best chance of escaping the attention of enemies. But it strongly dislikes being in the open for long, and soon will dig itself in again. For this operation it is admirably equipped. Working backward, it digs the apex of its pointed cone-shaped abdomen into the loose substrate with jerky movements, being aided in this by the rows of bristles set at a forward slant around its abdominal segments (fig. 11, left). It can disappear from sight in a few seconds. As soon as this happens, however, the sensation of sand on its flat head releases a head-jerking reflex. Since it continues to hurl away the sand above its head, a funnel of steadily increasing depth is formed. During this activity, the animal turns slowly around its own axis, a movement which results in an even distribution of sand all around so that the trap is equally accessible from all sides. The funnels built by *young* ant lions are very small. Those of large animals may reach ten centimeters in diameter. As soon as the pit has reached the desired size, its architect turns to the main object of its long larval existence: to watch and wait. Motionless, with wide-open jaws, it waits in ambush until a falling victim releases the snapping reflex. These two reflexes, the hurling reflex and the snapping reflex, are its main activities and the basis of its livelihood. Otherwise it moves but rarely; when it gets cold it just retires a little deeper into the soil.

I do not think we need to pity the ant lion because of the monotony of its existence. Boredom is, presumably, the doubtful privilege of creatures that know what amusement is.

The web of the garden spider

The homes of our garden spiders and those of many of their relations are very grand affairs: they are made of pure silk. Yet their owners do not have to pay for this luxury for each has her own private silk mill—a remarkable installation which almost completely fills a spider's abdomen. Essentially it consists of six pairs of glands capable of producing silken threads independently of each other. Each gland is connected by a duct to wartlike structures, known as "spinnerets," which are transformed vestigial legs at the rear of the abdomen (fig. 13). A spider about to spin a thread can choose among the secretions of these various glands what is most suitable for any given task. For though all glands secrete silk, which is a protein substance, the threads produced differ in consis-

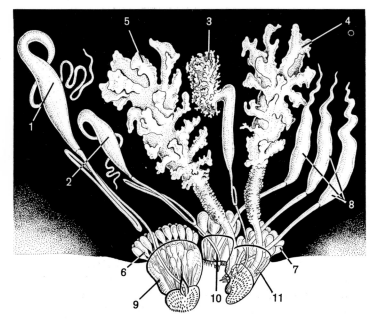

*Fig. 13. The spinning glands of a large tropical spider (*Nephila madagascariensis*). Only glands of the right side of the body are shown; complete system would be doubled. The bottle glands (1 and 2) provide the dry thread which the spider uses to move about the web, to drop to the ground, and to climb up again. The sticky thread is produced by the thread gland (3) making the basic thread, and the glue glands (4, 5) coating it (see pl. 15, p. 37). Gland 6 supplies an adhesive for fastening dry thread to a surface; gland 7, a multitude of fine threads to wrap up the prey; and gland 8, the material for the egg cocoon; 9, 10, and 11 are the anterior, middle, and posterior spinnerets of the right-hand side.*

tency and purpose. The whole system is immensely complex and involved—so much so that it could not be studied satisfactorily on an animal as small as our garden spider. However, in the body of *Nephila madagascariensis,* a tropical orb-making spider of such gigantic size that it can grasp a human hand with its outstretched legs, all details can be clearly recognized. Its spinning apparatus, which resembles that of our garden spider, is shown in figure 13, where a longitudinal section of the abdomen is presented and the glands of the right half, with the connecting ducts to the spinnerets carefully disentangled, are exposed. The caption explains the functions of the threads manufactured by the different glands.

The web of a female garden spider (only the females make large orbs) serves its owner both as a parlor and as a trap to catch flies and other airborne insects that cannot see the gossamer threads (fig. 14 and pl. 14, p. 37). Most of the time the spider sits in the center of her web; but a small hideout at its edge also forms part of her home, and because this is strengthened with leaves woven together, it provides a useful shelter at night or in times of bad weather.

The web must be sticky in order to serve its purpose; yet it is important that the spider herself should not be hindered by the stickiness whether she sits in the center of the orb or runs about on the strands. Therefore, the spider uses more than one kind of thread to construct

Fig. 14. A garden spider on her web, sitting in her lookout, from which spokes radiate to the frame of the orb. The sticky trap spiral lies between lookout and frame.

her web. The center of the orb, the "lookout" where the spider usually sits motionless waiting for prey, consists of dry threads produced by two pairs of bottle-shaped glands (fig. 13:1 and 2). From this central hub, other dry threads, originating from the same glands, radiate as "spokes" to the rim. The "trapping spiral," a sticky thread attached to the spokes, fills the space between hub and rim (fig. 14). It is produced by two different types of glands: the thread gland (fig. 13:3) provides the basic filament that during the process of extrusion is coated with a sticky substance secreted by the glue glands (fig. 13:4 and 5). Almost as soon as it is secreted, this sticky substance contracts and forms little beads (pl. 15, p. 37). This is the "birdlime" in which flies and other insects of medium size get hopelessly caught if they happen to touch the web. The spider herself, moving about her web, takes hold only of the dry threads and cleverly avoids touching the sticky spiral. This is made possible in the first place by the anatomy of her legs, which end in an arrangement of little claws and bristles (fig. 15), thereby enabling her to grip firmly the thin yet robust threads; and in the second place by the fact that the web is never vertical—as one will notice on close observation, it is always slightly inclined. The spider sits on the underside, or moves about in an upside-down hanging position, which makes it easier for her to avoid the sticky thread.

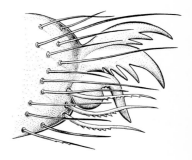

Fig. 15. The tip of a garden spider's foot, magnified.

It is obviously not an easy task to put oneself mentally into the place of a garden spider lying in wait for her prey. However, we know from numerous observations that her eight little eyes do not help her to perceive it. It is her extraordinarily well-developed sense of touch that is all-important. Even when she sits in her little shelter outside the web, she always keeps one of her front legs on one of the spokes or on a specially constructed "telegraph wire," for it is the vibration of the thread that betrays the presence of prey in the net and the tension of the spokes that tells her whether the catch is large or small. The exact whereabouts of her victim is also discovered in a similar manner. If the prey does not move, the spider, from her position in the center, locates it by plucking each thread in turn; because of the acuteness of her tactile sense, she does so at an amazing speed. This is of considerable importance to her, because insects of greater size and strength can often manage to struggle free despite the glue. Hence, the spider throws herself on her victim without delay, and secures it by swathing it in a mass of very fine silk threads from yet another set of glands (7 in fig. 13), turning it round and round all the while with deft movements of her legs. In no time at all she renders her victim completely helpless. Moreover, she bites it several times with her sharp jaws that are connected to poison glands—the injected poison kills a fly in a few minutes. The spider then releases the trussed victim from the web by severing the threads holding it, carries it to the lookout in the center, and hangs it up by a short thread. Like ant lions, spiders feed by injecting peptic juices into their prey and sucking them up later when all the nutrients are dissolved. The indigestible exterior skeletons are thrown out of the web.

Ant lions give the impression of being rather dull creatures. But no one would call spiders dim-witted. That they even possess a memory of sorts can be demonstrated by the following experiment. A spider has caught a fly, trussed it, and carried it to her lookout, but before she has sat down to her meal, the thread on which the prey is hanging is carefully cut by the experimenter so that the fly falls to the ground. The spider does not accept its loss lying down, but immediately embarks on a systematic search for her vanished meal. One by one she tests the spokes: finding no sign of a load, she even examines the place near the circumference where the fly had enmeshed itself in the first place. It takes a long time of fruitless searching before the spider calms down sufficiently to sit in wait for the next victim.

Spiders differ from insects in a number of ways. Most insects have wings, but all spiders are wingless. Insects have six legs, spiders have eight. Most insects hatch as larvae and pass through metamorphosis to become adults. Spiders, from the very beginning, look like their parents.

Even tiny young spiderlings fashion proper orbs which, though obviously not strong enough to hold, say, a big fly, might nevertheless catch a flying aphid. Without any prior guidance, they can carry out the complicated sequence of operations no less well than their mothers, and are even capable of taking advantage of local conditions by suitable adaptation.

How exactly does a spider fashion her web? It is no easy task to follow all the details of its construction, and I fear that the patience of my readers would be overtaxed by a full description. I shall, therefore, do no more than describe the main steps and explain the principles involved.

Let us assume our garden spider sits on the trunk of a tree. In order to be able to make a free-hanging web, her first step will have to be the erection of a kind of bridge to some other solid object. As we have shown, the rear end of her abdomen carries a set of cone-shaped spinnerets from which she can produce silken threads as from the finest nozzles imaginable. The spider lifts her abdomen and secretes a little silky substance, at the same time stretching the spinnerets, which are movable, as far away from each other as they will go. In this way the silky threads will form a kind of fan-shaped little sail, or kite, light enough to be carried on the slightest current of air. Next, she inclines the spinnerets toward each other so that the silken threads now emerging combine into one single strand, which on release is carried off at random behind its kite. If it does not settle anywhere, the spider hauls it back, eats it so as not to waste precious spinning substance, and tries again. If a thread comes into contact with a twig or some other solid object it will adhere to it; as soon as this happens the spider will fasten it at her own end, and the first bridge is ready. She immediately starts to cross it, but does so in a very peculiar fashion: she severs the thread by biting it through, but keeps hold of both ends with her fore and hind legs while her body is suspended between them. Running forward, she shoots out a fresh thread from behind and winds up the thread in front. By producing more thread than she rolls up, the thread is lengthened so that the bridge sags (fig. 16, a). When she has got halfway, she sticks the two ends together and lets herself drop to the ground (fig. 16, b), whereupon she takes a few steps sideways (i.e., in fig. 16, b, toward the observer or away from him) before securing her thread to the ground. The first three spokes of the future orb are now ready. The few sidewise steps serve to produce the slight incline so important for locomotion upon the hanging web.

The next step is the construction of the outer frame. It needs great powers of concentration and observation to follow how this is done. The spider climbs back to M, the future hub of the orb, and runs from there to A (fig. 16, c). As she releases a thread while doing so, that spoke is now doubled. She then retraces her steps, still paying out thread so that there are now three parallel strands; then she stops. Her next action may, at first, seem senseless: she sticks the newly spun thread 3 to thread 2 (fig. 16, c). However, we soon discover the reason for this:

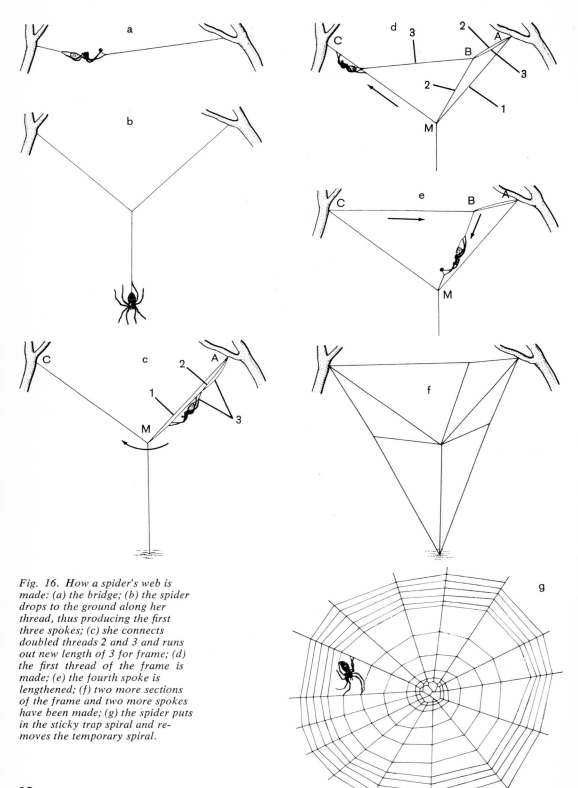

Fig. 16. How a spider's web is made: (a) the bridge; (b) the spider drops to the ground along her thread, thus producing the first three spokes; (c) she connects doubled threads 2 and 3 and runs out new length of 3 for frame; (d) the first thread of the frame is made; (e) the fourth spoke is lengthened; (f) two more sections of the frame and two more spokes have been made; (g) the spider puts in the sticky trap spiral and removes the temporary spiral.

as she runs (from the point B where thread 2 and 3 were joined, shown in fig. 16, d) via M to C, she reduces her output of silk; this tightens thread 3 and pulls thread 2 away from thread 1. Thread 3 is then fastened at C (fig. 16, d). This maneuver thus results in the formation of a new spoke (BM) and the first part of the outer frame. However, that part is not yet perfect—it is excessively taut and pulled away from the straight line between A and C by the fact that the spoke BM is too short. This is a fault that can be remedied: the spider runs back to B and proceeds to lengthen the thread between B and M by the technique already described, that is, by biting it through and replacing it gradually with a longer one, the animal itself forming a live bridge moving from B to M (fig. 16, e). As other sections of the frame are constructed in the same way, always in conjunction with new spokes, the structural pattern of the new orb emerges (fig. 16, f) though further spokes are still being added. Cross-threads linking the spokes together at the center form the foundation of the future lookout.

The most important element is still missing—the sticky thread to trap the insects. To lay this, the spider must first construct a temporary spiral from dry thread. This she does by walking around her web four to five times, climbing from spoke to spoke. As soon as the circumference is reached, she turns and, proceeding toward the center, weaves a much more tightly wound spiral of sticky thread in between the wider rings of the dry spiral. She moves from spoke to spoke, using the dry auxiliary thread, biting off those parts that become redundant as the work progresses (fig. 16, g). By the time the rays of the sun have induced the insect visitors to take to the wing, the web that she started early in the morning is completed.

It would be wrong to assume that from now on our spider can sit back and enjoy the fruits of her labor day after day. The glue thread does not stay sticky for very long, and therefore the web has to be renewed after a day or two. Only the frame, and less frequently the spokes, are used again. Sometimes rebuilding has to be started even earlier. When a lively bluebottle fly or a fat bumble-bee gets caught in the delicate web, the outcome may be devastation rather than a gorgeous feast. If this happens, the spider will start straight away to repair the damage, using once more the remarkable sense of touch in her legs to find out which part or parts require renovation or tightening.

Spiders that hunt in other ways

Different spiders—some twenty thousand species are known—construct different kinds of web. But there are also spiders that attack their prey as free-roving hunters, such as the wolf spiders and the tiny delicate jumping spiders. Their eyes are better developed than those of the web-spinning species. However, they too are spinners. Jumping spiders make loosely woven webs to provide shelter at night. And wolf spiders, which can often be observed moving about on the shores of lakes or on seaside beaches, attract people's interest at the time when they carry a conspicuous white silk cocoon under their abdomen; this cocoon holds the spider's eggs.

Other species hide their egg cocoons instead of carrying them about. Still others attach them without concealment to a leaf or some other surface. The dainty shape and snow-white color of the cocoon of the soil-inhabiting European tube spider *Agroeca brunnea* has prompted the name "fairy lamp" (pl. 16, p. 37). Its bulbous part, approximately one-half a centimeter in width, is attached by a kind of stalk to a blade of grass or the stem of some plant; it contains about fifty eggs. Its conspicuousness presents a danger, of course, but the cocoon does not stay white and shiny for long. The mother picks up small particles of soil or sand with her mouthparts and covers the tiny silky tent with a dark crust, using fine threads to fasten them securely. In this way, the shining little fairy lamp is quickly turned into a small lump of dirt that does not betray what is inside (pl. 17, p. 37).

One remarkable spider lives its entire life underwater and yet remains an air-breathing animal: the water spider *Argyroneta aquatica* is found in the stagnant or slow-moving waters of ditches and ponds all over Europe and Asia. This spider always carries her own supply of air about with her. It surrounds her abdomen and lower thorax like a silvery cloak and adheres tightly to the furry hairs of her body (pl. 18, p. 38). To keep it renewed, she need not rise to the surface more than about once a day. Her home is an air-filled underwater balloon held in place between aquatic plants and submerged branches by a network of threads spun in advance of its being filled. To fill it with air, the spider rises to the surface, traps an air bubble by crossing her hind legs over her upturned abdomen, and carries it down to her nest held between legs and abdomen. She then releases it under the web with a stripping movement of her hind legs. By repeating this a dozen or so times, she manages to fill a hemispherical divers' bell about two centimeters in diameter. Its regular shape may be slightly distorted where it is anchored to rigid twigs or roots. From her underwater nest, or else on hunting forays, she catches fresh-water isopods, aquatic insects, and other prey.

Some spiders inhabit tubes in the ground. Their homes are a strange contrast to these aquatic palaces or to the aerial abodes of garden spiders, but they are no less interesting. For a silk-lined tube with a close-fitting door is not a bad contraption. The family of trap-door spiders (Ctenizidae) lives mostly in the tropics and subtropics, but some species occur in the Mediterranean region, including *Nemesia cementaria,* which we shall take as our example. This spider burrows a tube-shaped sloping hole in a dry exposed bank (fig. 17), coats it with a mixture of soil particles and saliva, and thereafter lines it with silk threads, especially near the opening. A strong lid

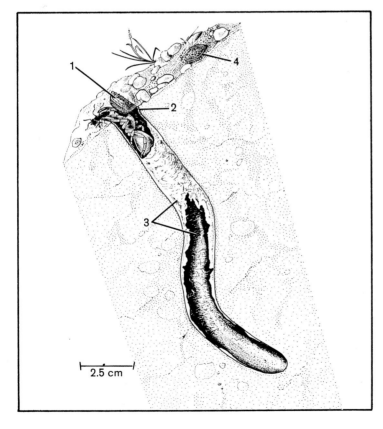

2.5 cm

Fig. 17. The tube of the trap-door spider (Nemesia cementaria) *on a dry slope. The spider, lying in wait for prey, whereby she keeps the lid (1) of the tube half open, has just caught an ant. (2) The hinge of the trap-door. (3) The tube and its silk lining partly cut open. (4) Closed lid of another tube is visible farther back.*

incorporating soil particles to give it weight is woven to cover the nest. This is fastened to the top edge of the opening by a strong silken hinge in such a manner that it can be easily opened but will close automatically by its own weight. The wide flange of this trap door is molded conically to ensure an exact fit to the mouth of the tube so that neither light nor rain water can get in. Its outer side is always camouflaged with soil from its immediate surroundings so that it blends with the environment and almost completely conceals the entrance to the tube (pl. 19 c, p. 38). The spiders, which can reach an age of ten years, stay in their dark tubes all their lives and never leave them. Even when they come out to pounce on a prey, they remain anchored to the entrance with the tips of their hind legs. However, by opening the lid wide and fastening it in this position with a pin, one can induce the spider to come out. It will appear at the entrance to investigate, and after trying in vain to shut the door from there, will come out into the open to do so (pl. 19 b—I am indebted to Dr. Friedrich Schremmer for these photographs). Strictly speaking, only the females are completely housebound. Once the males are mature, they leave their tubes at a certain time to look for a mate.

Trap-door spiders hunt by night and keep their front doors closed in daytime. When night draws close—which the spider knows by a built-in sense of time—it now and then lifts the trap door just a little to see whether it has gone dark yet. If so, the spider keeps the door half open and stretches its two front pairs of legs out of the tube (fig. 17). It may stay in this position for hours, motionless, waiting for insects. As soon as an insect, usually an ant, comes close, the spider pounces on it with lightning speed, seizes it, and pulls it into the tube; the trap door then closes automatically.

When the time of molting approaches, the lid is fastened down firmly with threads to afford safe protection for the short critical period when the old rigid armor is discarded and the new armor has not yet had time to harden. Before oviposition the door is sealed for a longer period, and the mother guards the egg cocoon at the bottom of the tube. The young spiderlings stay in the tube with their mother for at least a year, sometimes even longer. Looked at in the right spirit, such a dwelling with its silken hangings and ingenious door will appear more like a cozy home for a spider family than like a dark and dismal hole.

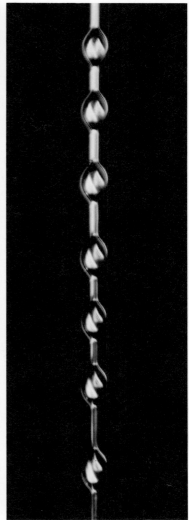

Plate 14. Spider's web in early morning. Repairs must be carried out every day; sometimes the whole web has to be renewed. (See p. 28.)

Plate 15 (upper right). Sticky trap thread from the web of the orb spider (Nephila), greatly magnified. The glue has contracted into droplets along the basic thread. (See fig. 13, p. 27; also p. 29.)

Plate 16. This little silken bell is called a "fairy lamp." A spider has fashioned it as a protective cover for her eggs. (See p. 34.)

Plate 17 (far right). The newly completed fairy lamp is camouflaged by its maker with little particles of soil, collected from the ground one by one. (See p. 34.)

Plate 18. *The dwelling place of a water spider: an air-filled bell anchored under water by a net of spider's threads. (See p. 34.)*

Plate 19 a (bottom left). *The trap door of the tube of* Nemesia cementaria *was opened extremely wide and fixed in that position with a pin (top of left and center frames). The spider comes out to investigate and tries in vain to close the trap door.*

Plate 19 b and c (center and right). *The spider has to come right out of her tube in order to free the lid and draw it shut. (See p. 36.)*

Plate 20. A caddis fly larva has attached small pebbles and snail shells to its casing. (See p. 43.)

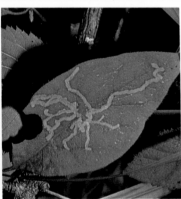

Plate 21. The mines of a micro-lepidopterous caterpillar inside a leaf. (See p. 46.)

Plate 22. Bagworm, crawling. (See p. 47.)

Plate 23. Bagworm, fastened to a stalk. (See p. 49.)

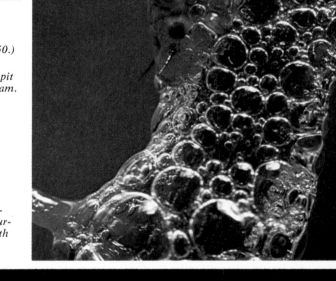

Plate 24. Because of their long stalks, the eggs of the lacewing (Chrysopa flava) are not very easily detected by ants. (See p. 50.)

Plate 25. Larva of the cuckoo spit insect, or spittle bug, making foam. (See p. 49.)

Plate 26. The digger wasp (Epibembex rostrata) excavates a burrow in the sand and stocks it with paralyzed flies as food for her larva. (See p. 51.)

Underwater nets

Spiders, of course, are not the only animals that produce silk. Many insects do, and silkworms have been cultivated for this purpose for a very long time. It is known from excavations that the Chinese used silk and had a highly developed weaving technique thirty-five hundred years ago. Before it was partly superseded by man-made substitutes, silk was a most important article of international commerce.

The spinning glands of spiders have their openings at the rear end of their bodies. Those of insects are modified salivary glands opening outward from the mouth; only larvae have them. On woodland walks one can often observe a small caterpillar "spitting out" a thin thread and using it like a rope to get down to earth from the canopy. In neglected orchards one often comes across groups of gregarious caterpillars that use their silk to make rough dwellings from tangled strands, or other species that weave urn-shaped bags to serve them as communal shelters. Many species, including the silkworms, fashion a cocoon in which to spend their pupal period. While they release the silk, they make a series of circular movements with their heads and the front parts of their bodies. The number of gyrations needed is great indeed, for a single cocoon contains three to four kilometers of thread! This also gives an idea of the fineness of the filament.

It is less generally known that certain insect larvae use their silk to make elaborate trap nets under water. In figure 18 is shown by way of example the trapping net made by the larva of a caddis fly.

Caddis flies are not really flies, but belong to the order of the Trichoptera, which is closely related to butterflies and moths. The winged adults are grayish brown and inconspicuous—they are easily taken for moths by the nonspecialist. They are often found near water, and their larvae, which look rather like caterpillars, are aquatic. Most of them construct quiver-shaped casings for their bodies (fig. 19, center, p. 44). But the larva that constructs the trapping net (fig. 18) can dispense with this type of armor. It finds excellent protection at the bottom of its funnel-shaped web, which it builds in sluggish streams and anchors to aquatic plants or branches. The running water keeps the funnel open and wafts all sorts of small organisms onto its walls for the larva to browse off at leisure.

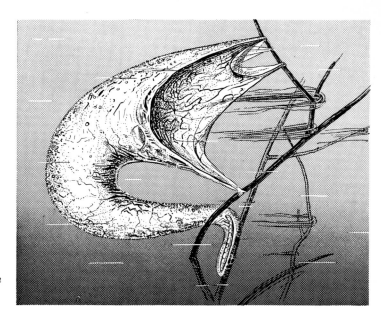

Fig. 18. A funnel-shaped fishing
net spun from silk threads. Its
maker, a larva of the caddis fly
(Neureclipsis bimaculata), stays in
the narrow part of the funnel.
About natural size.

BUILDING A HOME FOR ONESELF

The larvae of caddis flies

In the last decade of the nineteenth century, shady ponds
and other natural waters (soon, alas, to be swallowed up
by a relentlessly advancing civilization) were still quite
common on the outskirts of my native city. On the bank
of one such pool you might have come across a little boy,
flat on his tummy near the water's edge, gazing intently at
the tangle of aquatic weeds, decaying leaves, and broken
reeds covering the bottom of the shallow pond. He was
out to find something new and exciting for his aquariums,
something to add to his tadpoles, aquatic snails, and other
small creatures. With amazement he watched what had
seemed to him a twig or some other piece of debris suddenly start to move and slowly crawl away. He had to
look sharp to discover that this was, in fact, the casing
of an insect larva. Its head and legs could be seen protruding at one end of the protective sheath that hid its
soft abdomen; it carried the sheath about much as a snail
carries its shell.

In the last section, we discussed caddis fly larvae that
built themselves underwater fishing nets. Their way of
life is rather an exception in their circles. Many more

species of caddis fly larvae, or caddis "worms" as they are also called, build themselves similar quiver-shaped casings to live in. Almost as soon as the little larvae hatch from their eggs they start to weave dainty little tubes with material supplied by their salivary glands, a silk like that used by their cousins for the making of fishing nets, or the threads of silkworms and other caterpillars. Next, they strengthen this soft silky structure with more resistant material. Using their legs and sharp-edged jaws, they pick up such things as small bits of dead plant material, cut them to shape, and attach them to their tubes with silken threads. Work is carried out only at the upper end of the tube, that is, within reach of claws and legs, the abdomen remaining firmly inside (fig. 19 a, p. 44 and pl. 20, p. 39).

Some insect collectors specialize in these peculiar structures and enjoy their great variety just as much as a philatelist enjoys the variety of his stamps. For each species has a characteristic way of arranging its building materials. Some align little bits of broken twigs in parallel with the axis of the tube; others put them crossways; others still arrange them in the form of a spiral, or just join them together in a haphazard way. A species found in torrential streams in Japan attaches its casing to a stone by the broadened end of a curved "stalk" (fig. 21: 6, top row right, p. 48). Not only the "architectural style" varies between species; there are corresponding differences in the choice of the building materials employed. Some use dead vegetable matter, others small particles of sand or gravel; yet others attach to their casings the shells of tiny bivalves, or snails, occasionally including one that is still inhabited, thus causing its hapless inhabitant to die of slow starvation.

The favored material may not always be available, however, and the larvae may have to make do with what they can find locally. But this will ensure that the casings blend perfectly with the environment, and successfully camouflage their owners.

As a rule, the whole larval phase is spent in the same casing, though this has to be enlarged, of course, to provide room for its growing inmate. When the casing gets too tight, it tends to crumble at the lower end. However, some sheaths, made of very resistant material, are more durable and show the increase in size of the developing larva. A spiral construction like that of a snail's shell is the best way to conserve the older parts, and this design is adopted by members of the genus *Helicopsyche* for their peculiar little abode (fig. 21:5). It is largely a tropical genus, though some of its members have also been found in Lake Mendota in North America, and in southern Europe.

Caddis fly larvae are found in pools and lakes, in large and small rivers, and in running brooks. Obviously, they have to adapt their building materials to the nature of their aquatic environment. Lightweight casings made of bits of reed or leaves are suitable for stagnant water. Species inhabiting moving waters tend to prefer heavier stuff and frequently weigh down their cases by attaching larger pebbles.

The life led by the larvae in their silk-lined tubes is comfortable and sedate. They never hurry. Their portable homes are not made for rapid locomotion along the bottom, but since these larvae are not carnivores out to catch swift-moving prey, there is no need for speed. They usually feed on vegetable matter, live or dead, and this is plentiful enough where they live. When they are disturbed, they withdraw their heads and legs into their casings, and it is then almost impossible to dislodge them because their bodies are equipped with little protuberances which they press hard against the walls of the tube, and with

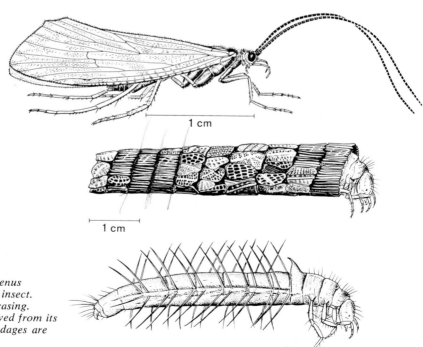

1 cm

1 cm

Fig. 19 a. Caddis fly (genus Phryganea). Top, adult insect. Center, the larva in its casing. Bottom, the larva removed from its casing. The long appendages are gills.

44

two small hooks at the rear which they use to take firm hold, resisting to the utmost. This is a very sound instinct, for under natural circumstances their soft bodies, once pulled out, would soon end in the stomach of a predator.

When the larva is fully grown, it pupates inside the case; but before doing so it fastens the case to the ground or to a plant; sometimes it weighs the case down with stones attached by threads (fig. 19 b, left). Occasionally the larva seeks a hiding place such as a crevice in the ground. It closes the opening of the case with a fine membrane, but leaves a slit to admit water (fig. 19 b, right) since pupae, too, need oxygen from the water to live.

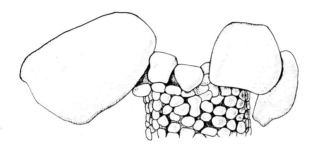

Once the necessary preparations to ensure a peaceful metamorphosis have been completed, the insect gradually changes from a larva to a winged adult. When it emerges from the pupa, it rises to the surface and takes to the air (fig. 19 a). Caddis flies sometimes attract attention by their large numbers and clumsy flight, but the beautifully constructed cases of their larvae are known only to few, and their connection with these fluttering swarms to fewer still.

Fig. 19 b. Larval casing from a species of caddis fly living in fast-moving water (Sericostoma personatum). Left, quite large pebbles are attached to it to weigh it down. Right, a view of the closing membrane of the casing showing the slit left for the water needed for breathing.

Larval dwellings of Microlepidoptera

Many people find it difficult to believe that the cases of caddis "worms" are sedulously sought by some collectors. But everybody knows that there are collectors with a passion for butterflies and moths. Most people will think of our beautifully colored butterflies and of the large, big-bellied moths that circle around outdoor lamps on a warm summer's night. They know little or nothing of the Microlepidoptera, though that large group of tiny moths embraces nearly a quarter of all known lepidopterous species. Nor are many collectors interested in them, for they are difficult to mount and to identify.

However, since these insects cross our paths not infrequently, we ought to know something more about them. Quite often, for instance, we observe on the leaves of roses and other plants some curious light meandering lines. They are evidence of the galleries made by the caterpillars of certain leaf-mining species of Microlepidoptera (pl. 21, p. 39)—animals so minute that they can hollow out a path inside a leaf without injuring the cells of its upper or lower epidermis. Since they eat only the tender interior cells—that is, the cells that contain the green pigment chlorophyll—their mines show as light lines on a green background, indicating where the colorless epidermis alone remains. The little larvae manage to combine the business of feeding with the construction of a tubelike dwelling, the roof and floor of which consists of the tough, untouched surface layers of the leaf.

Or take the worms occasionally found in apples or plums. They too are the caterpillars of two pretty species of small moths. The fruit serves them both as food and shelter, but unfortunately they spoil its flavor for us by the taste of their excreta.

The clothes moth is presumably the best-known member of the whole group. This household pest, once feared and hated, has lost much of its terror, thanks to our modern magicians of the chemical industry who provide us with mothproof woolen materials. It should be made quite plain that the drab-colored moths flying about in our rooms, which we so vehemently pursue and kill wherever we can, do no harm. Though they live for several weeks as winged adults, they never eat at all but live on the reserves built up during their larval past. Moreover, most of those we see are males, and a few more or less makes no difference. It is the caterpillars, feeding on animal hair or feathers, that do the damage. We ourselves would never eat hair or feathers, nor would most other animals, however ravenous, for they are completely indigestible. But the caterpillars of the clothes moths are an exception. Their peptic juice contains special enzymes capable of dissolving keratin, the horn substance of which hair and feathers are made. Since it consists of proteins, it is a valuable source of nutrients.

Because of their instinct to lay their eggs in the fur or plumage of dead animals, these moths perform a useful role as scavengers in nature. They are much less beneficial when they come indoors and deposit their eggs on our garments and upholstery. Their small caterpillars hatch after only a week or two and immediately embark on their work of destruction. They are sedentary by habit, and because they feed continuously in one spot, holes may appear in a single day. To make themselves comfortable they spin little silken tubes, covered with bits of wool or hair taken from the material or fur they sit on. These tubes are firmly attached to their substrate, but the larvae need only put their heads out to eat; when they have eaten everything within reach, they just lengthen their tubes a little and go on feeding close by. This explains why moth holes have such clean edges (fig. 20). Should the place chosen by its mother prove unsuitable or lacking in food supply, a caterpillar will leave its tube and crawl slowly away. When it finds a better pasture it will build another tube, but it will never match the larvae of the caddis fly in the exercise of this art.

There is, however, one family of Microlepidoptera that can stand up to the caddis flies, namely, the insignificant-looking bagworms, or case worms (Psychidae). More surprisingly still, the styles of these two groups of builders, whose paths never cross, have very great similarities. The male psychids are small, dark-colored moths with an adult lifespan of but two days. The females, which are wingless, remain sitting where they emerge from pupation, and in that same spot they attract the males (by scent) and lay their eggs. Immediately after hatching, the tiny caterpillars spin little silken tubes to which they attach plant or soil particles and which they never leave again. Like the caddis worms, they can carry their little homes about with them (pl. 22, p. 39), but are even slower in their movements. They are to be found on tree trunks, where they feed on lichen—though perhaps it would be more correct to say that they are *not* found, or hardly ever. It is difficult for even the most experienced collector to spot them, because they cover their "sleeping bags" with material from their immediate surroundings and this makes them virtually invisible—not only to man but also to the sharp eyes of hungry birds. It goes without saying that they, too, enlarge their tubes as they grow, each species doing so in its own characteristic style. A few examples will show how much their styles resemble the caddis fly larvae. Again we find wood particles arranged in longitudinal, transverse, or spiral patterns. In one species (*Apterona*), the casing itself is helical in the manner of a snail's shell—not unlike that of the caddis fly *Helicopsyche*. There is also a species from Nepal in which the casing is attached to a long "stalk" and sticks

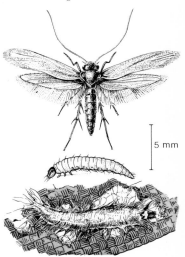

Fig. 20. Top, clothes moth. Center, caterpillar of the clothes moth. Bottom, the caterpillar in its little tube dwelling.

5 mm

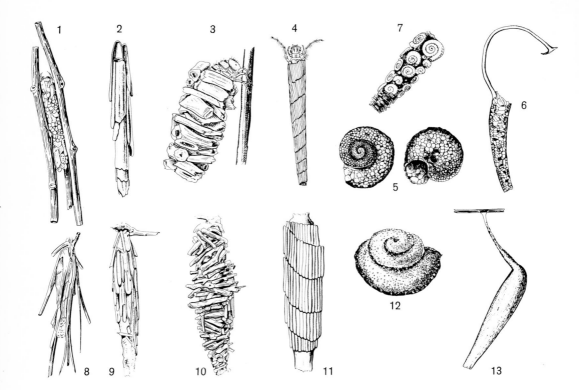

Fig. 21. Upper row, the casings of larvae of various aquatic caddis fly larvae (order Trichoptera). Bottom row, casings of the caterpillars of Microlepidoptera (bagworms, family Psychidae). Different species choose different materials and work in different styles. It is remarkable how often similar types occur in both groups of insects though they are not closely related. The caddis flies: 1, Anabolia sp.; 2, Grammotaulius nitidus; 3, Limnophilus flavicornis; 4, Triaenodes sp.; 5, Helicopsyche sperata; 6, Kitagamia montana; 7, Limnophilus flavicornis. The bagworms: 8, Basicladus tracys; 9, Oiketicus townsendi; 10, Oiketicus platensis; 11, unidentified tropical species; 12, Apterona sp.; 13, Metisa sp.

out at an angle (fig. 21, bottom row). Unlike the caddis worms, the bagworms never use small pebbles, presumably because stones are not easy to come by among the lichens and leaves where they live. Another reason may be that stones are only suitable as building material for dwellers in water because the water's buoyancy carries part of the heavy load.

It is strange enough that the females of these moths have lost their wings and have to wait for a male to find them wherever they happen to be. Stranger still is the fact that some species can do without males entirely—their eggs can develop without fertilization. This phenomenon of parthenogenesis (from the Greek *parthenos*, "virgin," and *genesis*, "creation") occurs in a number of animals and is well known to zoologists. But I myself had never heard of it when I started my insect collection as a boy, and for me it was a most exciting discovery. I did, of course, breed butterflies and moths from caterpillars since this is the best way of obtaining perfect specimens. One day I found a psychid caterpillar with its case and put it into a breeding box. It never left its casing—psychid larvae never do—but pupated inside, eventually turning into a wingless female without my noticing. Obviously, this caged female could not have been visited by a male. But when next I inspected the breeding box, I found the maternal casing surrounded by more than a dozen tiny, perfectly constructed larval casings containing her progeny. The mother herself remained invisible. She had stayed in her bag, as psychids often do, and had laid her eggs without bothering about male cooperation. But the baby caterpillars had moved out as soon as they were hatched and had built their own little homes without delay (fig. 22 and pls. 22 and 23, p. 39).

The cuckoo spit insect

The refuge constructed by the froghopper known as the spittle bug, or cuckoo spit insect, is of a very different nature. Adult froghoppers are easily recognized by their wings, which are shaped somewhat like a gabled roof (fig. 23, bottom), and by their marvelous prowess in jumping. They may suddenly jump on one's hand as one is walking in a meadow, and when gently touched from behind will bound away with a mighty leap, never to be found again. To unspoilt children who still enjoy simple pleasures, they are a delight. Their larvae (fig. 23, upper right) spend their lives in meadows or hedgerows, sitting on the leaves of willows and other leafy plants which they pierce with their sucking proboscises to feed on plant juices. They are not at all easy to see because they are surrounded by a protective ball of froth, the "cuckoo spit" (fig. 23, left) which, though conspicuous enough, does not look inviting to a predator. For instance, the ants, those formidable hunters of insects, leave the succulent froghopper larvae alone as long as they are covered with foam. But if the protection is removed, the delicate larvae soon fall victim to predators.

How is the cuckoo spit made? The sucking larva receives an abundance of water from the plant juices it feeds on; excreting the surplus through the anus, it virtually sits in a little pool of plant juice. This it turns into froth by blowing exhaled air into it (pl. 25, p. 40). To breathe in, the larva lifts the tip of its abdomen out of the fluid. A groove on the underside leads air to the openings of the breathing organs, the tracheae, a system of very fine tubes ramifying all over the insect's body. The air breathed out through these holes, or spiracles, produces the bubbles. Of course, the trouble with air bubbles in water is that they don't last. The sparkling foam of a

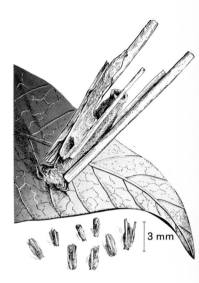

Fig. 22. A female bagworm had pupated inside her casing and had died in it after laying her eggs. The eggs, though not fertilized, developed; the young caterpillars emerged and immediately built their own diminutive casings.

3 mm

49

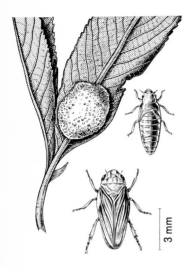

Fig. 23. Top left, cuckoo spit on a willow leaf. Top right, its originator, a froghopper larva. Below, the froghopper after metamorphosis, no longer living in a foam cover.

3 mm

waterfall bursts as it falls. To make a lasting lather, a substance like soap or detergent that modifies the tension of the membrane surrounding each bubble has to be added to the water. How the spittle bugs do this they have not revealed to us. But it has been found that a substance secreted by their kidney tubes has a stabilizing effect on foam.

This simple yet effective method of protection is not the exclusive invention of these insects. We shall meet it again in certain vertebrate animals.

BUILDING TO PROTECT THE OFFSPRING

Animals as a rule face fewer dangers as adults than in their juvenile stages. One way of making sure that despite enormous early losses individuals will survive long enough to perpetuate the species is the production of very large numbers of offspring. It is an effective if primitive method resorted to by many species of animals. The methods of insects are rather more sophisticated. They seek to lay their eggs in places where their larvae will find protection and suitable environment.

A simple way of doing this has been developed by the lacewings, which belong to the order of Neuroptera ("Nerve-wings"). The fully developed winged insects can often be found sitting on leaves, but are rarely noticed. Their wings, which they carry raised above their slender green bodies in a rooflike fashion, are large, green, diaphanous, and iridescent; the delicate tracery of their veins is a major characteristic of Neuroptera. When the female is about to lay an egg, she first secretes from a gland in her abdomen a viscous thread that quickly hardens into a kind of stalk. Usually the eggs rest upon these stalks a short distance from one another on the surface of a leaf. Sometimes they are deposited close to each other, and then their stalks may become entwined (pl. 24, p. 40).

In some highly developed insects, such as many bees and wasps, the females ensure the survival of their brood by skillful and dedicated building activities. "Building" in this sense may involve no more than simply making a hole in the ground, though the care of the progeny may be considerably more elaborate. Often, however, the structures themselves fascinate the observer by their ingenuity and their remarkable forms, which may differ fundamentally from species to species.

At the mention of wasps and bees, most people will probably think of the yellow jackets that swarm around fruit stalls and spoil our pleasure in outdoor picnics in summer and autumn, or of the providers of honey in the beekeepers' hives. Few will be aware of the existence of a great number of wasp and bee species that live solitary lives and remain largely unnoticed. After mating, the males stay about for a time, visiting flowers and sipping nectar. They take no part in the raising of the brood. The care of the next generation is left entirely to the females.

Digger wasps

As our first example we shall look at the family of digger wasps (Sphecidae). Like all other wasps and bees, as well as many other insects, they belong to the order of Hymenoptera ("Membrane-wings"). Some are of impressive size, but mostly they are medium-to-small insects showing the typical yellow and black stripes of yellow jackets; some are black and inconspicuous. The females work extremely hard. For each individual egg they construct a separate nest cavity either in the ground, or in decayed wood, or perhaps inside the stems of plants, according to the species. Sometimes their nests are scattered, sometimes aligned in rows, and sometimes arranged in clusters, but each cavity is a nursery for one larva only. These larvae live on meat, and this is provided by the mother wasps in a highly sophisticated, not to say shocking, manner. Each digger wasp specializes in a particular kind of prey. Some very small ones hunt aphids, while larger species prey on flies, caterpillars, grasshoppers, cicadas, or the like. Now if the wasps were to kill their victims, their flesh would most probably decay or be attacked by mold in the few weeks that must pass before a larva hatches from an egg that has yet to be laid and completes its development. However, this does *not* happen. For the prey is not killed but is paralyzed by stings; helplessly it must suffer being dragged into the nest and, later, being eaten alive by the hungry larva.

One such digger wasp, *Bembex (Epibembex) rostrata,* an inhabitant of the seashore and sand dunes, is shown in plate 26 (p. 40), digging a burrow into the sandy ground, throwing the excavated sand far back behind her. She widens the end of the burrow, forming a chamber which she provisions with paralyzed flies for her larva to feed on.

The Dutch ethologist G. P. Baerends has made a thorough study of another species, *Ammophila adriaansei* (the "sand wasp," or thread-waisted digger wasp), on a heath in his native country. This wasp, which is quite large (its body measures two centimeters), digs a vertical shaft in the sand and widens it at the bottom into a lateral chamber. For this she uses her mandibles and front legs, which carry a set of bristles making them highly suitable for the task. She then carries away the excavated sand by holding it between her "chin" and her thorax, and scatters it in the surrounding area. In favorable weather, she excavates a nest in about an hour and proceeds at once to close it again temporarily. Were she to use loose sand for the purpose, it would fall through the shaft to the nest chamber below. So she searches instead for a small pebble of a size that will just fit into the shaft (pl. 30, p. 58). She uses her wide-opened jaws as a measure, since it was their span that determined the width of the shaft in the first place. Nevertheless, quite a number of stones that cannot be wedged properly often have to be discarded after trial before one is found that really fits. Thereafter, the filling-in of the rest of the tube with sand and gravel is no problem.

After she has thus protected her nest against unwelcome visitors or even permanent intruders, she hunts in the neighborhood for a caterpillar. She stings it several times to paralyze it with her poison, and with great effort carries it to her nest (fig. 24 a), sometimes over considerable distances—up to forty meters or more. Her sense of direction, which guides her back to the nest area, is amazing. Once there, she recognizes the exact location of the closed nest opening, which is almost invisible, by the position of plants and other landmarks in its immediate neighborhood, and there she deposits her burden. She then reopens the shaft, taking care to set aside the well-fitting little "coping" stone, for she will need this useful piece again more than once. Next, she picks up the caterpillar and, walking backward, drags it into the burrow

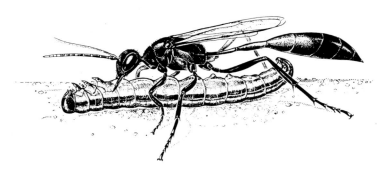

Fig. 24 a. The wasp carries a caterpillar which she has paralyzed by stings. Length of wasp, 2 cm.

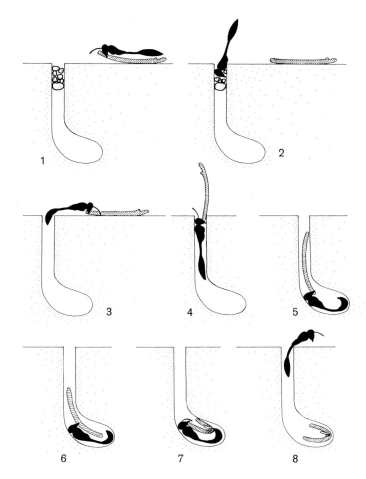

Fig. 24 b. The caterpillar is deposited on the ground (1). The nest, which is only provisionally closed, is opened (2) and the caterpillar is pulled into it (3–6). After she has laid an egg (7), the wasp leaves the nest (8), and closes it once more. Diagrammatic sketch (wasp shown without legs or wings).

where she deposits an egg on it (fig. 24 b), and fills in the nest shaft once more. Then she embarks on the construction of another nest. In contrast to the habits of most digger wasps, the female of this species cares for her brood even after the egg-laying stage. Watching her doing so, one might take her for a sentient being. Every morning she makes the rounds of those of her nests she has not yet finally sealed; these may be as many as two or three. She opens them and inspects their interiors. If the larva has not yet hatched, she just fills in the shaft again. But if it has emerged from the egg and is busy eating, its mother will catch another caterpillar or two to replenish the larder. When at the next inspection the larva has grown to a certain size, she will provide it with another six or seven caterpillars. This second restocking operation is the last. She then closes the nest definitively and does not return to it again.

This final closure is done with much greater care than the previous provisional fillings-in. After the coping stone has been put into position at the bottom of the shaft, the sand in the remainder of the hole is pressed down with the head. Wasps have frequently been observed to use a stone held between their mandibles to tamp down the sand of the top layer—one of the rare instances of the use of tools by animals. The familiarity with their coping stones acquired by the wasps in the course of their repeated temporary closure operations may have contributed to this remarkable achievement.

One would be inclined to think that the behavior of the wasp is both rational and sensible. Yet the whole sequence of actions is performed not by reasoning or insight, but entirely by instinct. This fact becomes clear when the activities of such a wasp are explored more closely with the aid of experiments. Whether or not further caterpillars are to be added to a burrow, and how many, is unalterably decided by the food situation encountered by the wasp on her first morning round of inspection to all nests that are not permanently closed. If an experimenter adds further caterpillars after her first visit, this action will not stop her from carrying in, quite senselessly, the predetermined number of caterpillars. Or should the experimenter move a paralyzed caterpillar from outside the nest entrance while the wasp is busy reopening the shaft, the bewildered insect will hunt around for the vanished prey and drag it back to the entrance. But instead of pulling it down right away, she will deposit it once more near the shaft and start digging in it although it is already open—that is to say, she will perform the action which under *normal* circumstances invariably follows on the putting down of the prey. Only then does she take hold of it to pull it down into the burrow. If the prey has been spirited away a second time, she will again search for it, return it to the nest entrance, deposit it, and start digging in the open shaft, as before. This kind of sequence might be repeated twenty times or more, for the wasp carries out a chain of instinctive actions without any intervention of thought or deliberation. The completion of one operation triggers off the start of the next and so on, each action following on the preceding one in a sequence that has remained unchanged through countless generations.

The fully grown larvae spend the winter in their nests and pupate in spring. The adults emerge in summer and

dig their way to the light, the males to lead a "life of leisure" and the females a "life of toil," building their nests
and caring for their brood as their mothers had done
before them.

True wasps

The true wasps (Vespidae) differ from other wasps by
their habit of folding their front wings lengthwise when
at rest—something that few insects do. Many solitary
species, and all social wasps, belong to this family.

A solitary true wasp.   One example of a solitary true
wasp to be discussed here is the potter wasp (*Eumenes*).
Whereas the digger wasps, including *Bembex* and *Ammophila,* excavate their nest chambers, the potter wasps fashion delicate little pots of clay. They attach them to a plant
or board singly or in groups, or hide them under the
loose bark of a tree (pl. 27, p. 57). To do so, the female
potter collects her material from a patch of clay soil. If
she finds it too dry for her purpose, she will moisten it by
spitting on it water she fetched and brought along in her
stomach. She will then scoop up small amounts of clay
and form them into little pellets with the help of her mandibles and front legs. These legs are saber-shaped and
make excellent tools for the job. Holding the pellets between head and thorax, the insect carries them to her
building site. Using once more her legs and mandibles,
she turns them into flat narrow strips and, adding strip
to strip, builds up a small hollow sphere with a narrow
neck and opening at the top, like a bottle (fig. 25 and
pl. 29, p. 57). It has been suggested that these nesting
chambers served the American Indians as models for the
shape of their clay jars.

When the structure has reached this stage, the wasp
goes hunting and collects a number of beetle larvae or
caterpillars, which she paralyzes. As she herself can no
longer get through the narrow opening, she presses and
squeezes them through the hole from the outside (pl. 28,
p. 57). When the nest is stocked to her satisfaction, she
inserts the tip of her abdomen into the opening and lays
an egg, which she attaches to the top of the vessel in the
following somewhat unusual manner. During the extrusion of the egg, she secretes a liquid which quickly hardens into a thread. On this thread the egg then hangs in
between the paralyzed larvae at the bottom, so that the
wasp grub can start feeding immediately after it has

hatched (fig. 25). As soon as the egg is laid, the wasp closes the opening with a final pellet of clay and takes no further interest in the nest (pl. 29, p. 57).

Plates 31a and 31b (p. 58) show the nest of another solitary true wasp of the genus *Oplomerus* (formerly *Odynerus*) that digs a burrow into a steep slope or mud wall. She mixes the material she has excavated with some of her saliva to build a curious tube of considerable length (up to fifteen centimeters), which hangs down from the mouth of the burrow (pl. 31a). Finally, she stocks its deep cavity with paralyzed beetle larvae that look like sawfly larvae and have often been mistaken for them (pl. 31b). The same kind of tube-shaped exterior additions to nesting cavities are also found with some solitary bees nesting in mud walls. Opinions as to their significance differ considerably, and nothing is known for certain. It has been observed, however, that these wasps use portions of the clay from the tube at a later stage to close the nest after it has been provisioned and completed.

Social true wasps. Most people are familiar with the imposing nests of paper wasps, a group of true wasps with social habits of which the largest European representative is the hornet *Vespa crabro* (similar to the American *Vespa maculata*). What a contrast these structures present to the simple, elegant little vessels of the potter wasps! There is, however, an important similarity: here, too, the foundation of the nest is accomplished by a single female, fertilized the previous year, which has survived the winter by hibernating in a sheltered spot. She becomes the queen of the new colony. At first, this queen wasp has to work entirely unaided, like the potter wasp. Without help she builds a home, lays eggs, collects food, and tends her first brood from which her future helpers will emerge. The eventual size of the nest is made possible through the lightness of the building material. This unique substance is a kind of paper made of small particles of wood shaved off by the wasp with her mandibles from

Plate 27. Five nests of the potter wasp on the inside of loose tree bark. Left, a wasp emerging. The cells at the right are no longer occupied. The emergence holes are clearly visible. (See pp. 55, 64.)

Plate 28. A potter wasp pushing a paralyzed caterpillar into the clay jar. (See p. 55.)

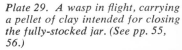

Plate 29. A wasp in flight, carrying a pellet of clay intended for closing the fully-stocked jar. (See pp. 55, 56.)

Plate 30. The sand wasp (thread-waisted digger wasp) tests the fit of her coping stone in the opening of her tube. (See p. 52.)

Plate 31 a. A solitary wasp has dug a nesting tube into a clay bank and, in addition, built an exterior tube which hangs down from the entrance. That tube consists of the excavated clay mixed with saliva. (See p. 56.)

Plate 31 b. Cross section. At the bottom of the tube, in a widened cell, a number of paralyzed beetle larvae are stored as food for the larvae of the wasp. (See p. 56.)

Plate 32 a. A young wasps' nest.

Plate 32 b. Front nest cover re-moved. The comb is attached to a support above by a stem; it will soon be enlarged and others will follow below. Flight hole at the bottom. (See p. 61.)

 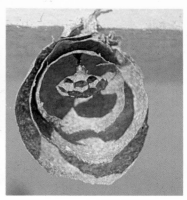

Plate 33. A hornets' nest (Vespa crabro) that has reached its full size. Most of outer nest cover has been removed. The horizontal combs are attached to each other by columns. (See p. 61.)

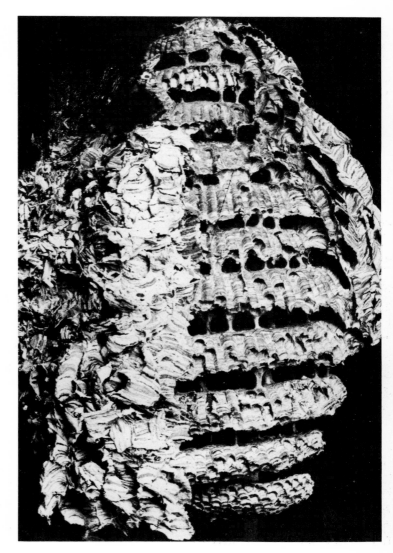

Plate 34 (right). Nest of the field wasp Polistes. *The single, coverless comb is attached to a stalk by a stem. (See p. 65.)*

Plate 35 (below). Wasps' comb with brood. Before the larvae pupate, they spin a lid over their cell. At the top, the queen. (See p. 62.)

Plate 36. A nest of the field wasp seen from above, with a queen sitting on it. (See p. 65.)

60

wooden beams, posts, boards, etc. Saliva is added to give it cohesion. Man has learned the art of making paper from wood fiber and a binding material by observing the wasps.

The vespine queen tends to choose a beam in a roof space or a similar sheltered spot for her nest. She uses some of the rapidly hardening paper pulp to make a kind of stem from which she suspends a very small comb. Whereas the wax combs of honeybees are vertical and carry cells on both sides. the combs of wasps are, as a rule, horizontal and have cells only on the underside. Both bees and wasps build hexagonal cells designed for one larva. This central comb, which will house the brood, is then surrounded with a multiple protective "envelope" of layers of paper. Only a flight hole is left open at the bottom (pl. 32 a and 32 b, p. 59).

The nest does not long remain so small and simple a structure. It is soon extended into a multistoried building. We humans start our houses at the bottom—these builders add story after story from the top downward, suspending each new comb from the one above by columnar supports. They also extend the combs laterally. As the interior structure expands, the outer walls, too, must be enlarged. Parts that are too close to the center are pulled down, and new ones, providing more space, are added. In this way, very large nests may develop in a single summer. One such nest (a hornets') is shown in plate 33 (p. 59). Hornets can sometimes be observed building the outer walls of their nests. They bring along one small pellet of paper pulp after another. Walking backward, they stretch the pellets into strips with their mandibles, much as the potter wasps do with their clay, and add the strips to the growing envelope, a method which often shows up clearly on the finished nest. The stem from which the nest is suspended and the columnar supports are made of the same material as the thin, almost flimsy, outer cover. Considerable load-bearing strength is achieved by aligning all wood fibers longitudinally—just as the tendons of muscles derive their immense toughness from the fact that all the fibers of connective tissue are aligned parallel to each other in the direction of stress.

The color of wasps' nests depends on the color of the wood used. Frequently they are gray (pl. 32 a, p. 59) because the wasps tend to collect their wood from the gray weathered surface of telegraph poles, fence posts, or boards. They can often be observed doing this. The reddish tint of hornets' nests is due to a preference for decayed oak wood.

The cells of the wasps serve exclusively for the raising of the brood. One egg is deposited in each cell, and the larva hatching from it is fed mouth to mouth. Wasps are carnivores. Although they enjoy sipping nectar and nibbling sweet fruits when they have a chance, the principal food of the European wasps, and the entire diet of their larvae, consists of other insects that they attack, reduce to pulp, and carry to the nest in the form of small pellets. When they are hungry, the larvae beg for food by making scratching noises on the cell walls. After three weeks they are fully grown and pupate, first closing their cells by spinning a membrane (pl. 35, p. 60). After another three weeks or so, they emerge as winged adults.

Though all the adults that come out of the cells in the earlier part of the season are females, they are not "queens" like their mother. They are much smaller, and their ovaries are but poorly developed. These workers, as they are called, now take over the jobs of building, foraging, and tending the brood, leaving the queen free to attend the more assiduously to her task of egg-laying. A *wasp community,* built on the principle of division of labor, has come into being.

This community is very comfortably housed. It even possesses a heating system. In the area of the breeding combs the temperature is kept constant at about 30° C. (86° F.) by a special group of workers. These act as living "heaters" by engaging in intensive muscular activity, contracting and stretching their abdomens in rapid succession. To a certain extent, the larvae themselves also contribute to the generation of heat by movement. The multiple paper envelope together with the air trapped between its layers forms an excellent insulating wall and prevents the loss of heat to the outside air. But should it get too hot in the nest on warm days, the wasps carry in water to moisten the cells, which are thus cooled by evaporation. No thermometer is needed. Their temperature sense tells them exactly when to heat and when to air-condition.

Not all wasps' nests hang from roof beams or branches where they can be easily seen. Right in the middle of a grassy meadow, one suddenly may find oneself attacked by a swarm of wasps because one has accidentally stepped on a "wasp hole." Such a ground nest is often attached to the roof of a small animal's burrow. When the nest grows, the original cavity becomes too small and is widened by the workers. They loosen soil particles and small stones and carry them away. Should a stone be too large and heavy to be shifted, the wasps remove the soil from under it, causing it to drop lower, and make room for the expansion of the nest in this way. Quite often there is a sizeable collection of stones at the bottom of the nest cavity, tangible evidence to a remarkable effort on the part of the small civil engineers (fig. 26). The choice of a nesting site—above or below ground—does not depend on the species. In many species, a queen wasp makes the choice of either a hole in the ground, a roof joist, or some other place high up in the air as a site for the foundation of her colony. This flexibility of the building instinct in certain species may have contributed to their wide distribution and to their dominant role in the insect kingdom.

In tropical regions, especially on the American continent, we frequently find a group of true wasps that have developed a different architectural style. In the genus *Polybia* and her relations, the outer cover of the nest is much stronger and tougher than the delicate paper envelopes of wasps' nests already discussed. It is often oblong, or tube-shaped, and is given added stability by the fact that the combs are firmly attached to the nest cover all around. This has the drawback, of course, that there is no room left between combs and cover for the wasps to move about from story to story as in the nests

Fig. 26. Subterranean wasps' nest. The cavity and the flight channel are shown open. At the bottom of cavity and in passage are stones which the wasps have been unable to carry away because of their weight.

of hornets and their close relations. Instead, *Polybia* and her relations build a communications shaft by leaving a hole in the center of each comb (pl. 38, p. 77). In tropical forests, these nests can often be seen dangling from the branches of tall trees like large sausages. They are strong enough to withstand exposure to the elements for several decades and thus ensure a long life to the wasp colonies inhabiting them.

In tropical climates, many kinds of plants and animals have developed a profusion of remarkable shapes and forms. In the same way, the tropics appear to have had a stimulating influence on the creative inventiveness of true wasps.

Plate 37 a and b (p. 77) illustrates the nest of the wasp *Polybia emaciata* from Colombia. This nest, which Professor Friedrich Schremmer detached from the branches of a shrub some two meters from the ground, is not made of paper. Like the solitary potter wasps (pl. 27, p. 57), these wasps work with clay and, like them, they are superb craftsmen. The outer cover consists of a mortar made from clay and sand. The combs and their hexagonal cells are modeled from a pure, very fine clay. The topmost comb is attached to a branch by a stem; a further support connects it to the domed top of the mortar cover, but the lower combs are not joined together by supports as those of hornets and their relations. They are attached to the sides of the outer cover. In this respect, the nest resembles that of *Chartergus* (pl. 38, p. 77). However, it differs from it by having no central flight shaft. Instead, a gap is left between combs and cover in the region of the flight hole, allowing the wasps to move from comb to comb. The nest is seven centimeters in diameter. Its builders are small and slender (about one centimeter long). In plate 37 a (p. 77) one of their number is shown sitting in the flight hole and another just above the entrance.

Polybia singularis has found yet another solution to the problem of internal communications (pl. 39, p. 77). The flight hole is a long slit easily accessible from most combs. These wasps, which are no bigger than the builders of the mortar nest (pl. 37 a), are capable of constructing imposing edifices. The one shown in plate 39 measures about thirty by fifteen centimeters and weighs exactly 1350 grams. The nest, which originates from the Upper Amazon region, is modeled from a very fine clay. Schremmer describes it aptly as a ceramic nest.

Some paper nests of tropical true wasps also show very

interesting peculiarities. *Metapolybia pediculata* attaches a small comb, circular or oval in shape, to the stem of a tree, a wall, or possibly a board. Suspended above it about one-half a centimeter from the cell openings is a roof of wasp paper supported by a circular wall built around the nest. The comb never touches this superstructure, the flight hole of which is in the lateral wall. The most remarkable feature of this very delicate cover is the fact that both in the roof and in the lateral wall there are many closely spaced tiny windows, measuring about one to three millimeters in diameter. The "window panes" consist of hardened saliva, a substance which we have already seen the wasps use as binder in papermaking. Part of the roof of such a nest, which Professor Schremmer brought back from Panama, lies on my desk in front of me. I should dearly like to know why this species provides illumination for the interior of its nest. It is probable, however, that the purpose of this construction is not so much interior illumination as camouflage. Several observers have noticed that the little windows break up the nest surface optically so that it becomes almost indistinguishable from the lichen-covered stones or the tree bark of its normal surroundings.

I have described the nests of highly organized social wasps immediately after those of solitary species. Obviously, a social organization must have developed gradually in the course of phylogenetic evolution and many transitional stages must have existed, but few can be found today. However, there is one kind of paper wasp (*Polistes*) whose nest may be looked upon as such an intermediate stage. It consists of a single small comb, without any kind of exterior cover, which is attached to a stone, a board, or the stalk of a plant (pls. 34 and 36, p. 60). Should it rain, the rain water will be sucked up by the wasps and carried away. When it is very hot, they fetch water to cool the comb. Since neither burrow nor outer cover shelters the little colony and its brood, its defense and protection is entirely the responsibility of the inhabitants.

A single queen, having survived the winter, starts building in spring. Other queens of the same species may join her and may participate both in building and in egg-laying. However, a hierarchy soon develops in an unexplained manner. The first female lays the largest number of eggs and devours those initially laid by the others. These later arrivals are thereby degraded to the status of worker wasps, or auxiliary females. After the first brood has emerged, the reigning queen drives all the other queens away. There is something oddly thought-provoking in finding already developed in a wasp state such methods of creating a central power.

Even more primitive transitions from solitary to social behavior have been reported from certain tropical Vespidae. I shall not go into these here, however. I shall have to take leave of the social wasps and turn once more to the structures of solitary Hymenoptera. This time these belong to the family of bees.

Solitary bees

The family of bees comprises more than twenty thousand different species, but few of them are social insects in the manner of the bumblebees and the honeybees. The great majority live alone—they are "solitary" bees, and many are not recognizable as bees at all except by experts. They vary greatly in size and appearance. Some are tiny insects, no more than two millimeters long, while others, measuring nearly four centimeters, appear giants in comparison. Some are nearly hairless, others furry. Many delight the observer by their varied patterns and attractive coloration.

Bees differ from the predacious wasps in one important particular: they are strictly vegetarian, feeding themselves and their brood on pollen and nectar. This habit endears them to kindhearted people, for they do not destroy in order to live. Moreover, their mode of feeding makes them an important life-enhancing force in nature's household. Flitting from flower to flower, they unconsciously act as plant breeders. By transferring pollen from one blossom to the next, they bring about fertilization and promote the setting of fruits and seed. Mutual adaptation between bees and flowers over millions of years has been largely responsible for the present advanced development of the fragrance of flowers and the splendor of their colors. For the greater the flowers' appeal to the senses of smell and vision, the easier it is for the insects to find them, and the better the chance of their pollination and propagation.

The nests built by solitary bees to protect their brood are immensely varied. Some look for a suitable cavity in the hollow stalk of a plant or in some piece of wood. Some dig burrows into the ground very much like digger wasps, but hollow out a series of lateral breeding chambers, one for each larva. Others practice masonry and build with mortar made from sand and saliva, and yet others are in the tailoring business, cutting out leaves according to a pattern and fashioning the pieces into small, thimble-shaped cups, but using neither needle nor thread. The homes of some bees look almost as if they were the work of architects simply bursting with ideas and using all kinds of techniques, with the all-important difference that the bees, guided by their marvelous instincts, can do it all without thinking.

We shall now look at a few examples of bees' nests. But it should be remembered that these are just a small selection from the great variety of existing structural types and that even within a single species there may be variants which we cannot describe.

The plasterer, or gum, bees (*Colletes*) and the diminutive yellow-faced bees (*Hylaeus*) are among the most primitive forms. These "ancestral" bees have certain similarities with the digger wasps from which, phylogenetically, the bees are descended. Their primitive characteristics include the absence of a long tongue adapted to the sucking of nectar from deep-cupped flowers and the lack of special equipment for gathering pollen. They neither possess the peculiar structure of the hind legs found in honeybees and in many solitary bees for the transport of pollen in "baskets," nor the furry brush on the belly which other species use for the purpose. Their method is to swallow the pollen and regurgitate it, mixed with nectar, into their breeding chambers as food for their brood. They build their nests in hollow twigs, holes in the ground, or other cavities. But their building activity consists only in lining these places with a wallpaper made of an oral secretion. This is a viscous fluid that soon hardens into a waterproof film like cellophane, which prevents the nectar from soaking into the ground and at the same time protects the contents of the chamber from dampness and mold.

The food prepared for the larvae of most solitary bees is a mixture of pollen and nectar known as "bee-bread" which, because of its high pollen content, can be stored in the brood chambers as a solid cake. In figure 27 a nest is shown that a hole-nesting species (*Heriades*) had built in an abandoned insect gallery inside a piece of wood. The cells are arranged in a line. The larva in the first (topmost) cell has nearly finished her ration of bee-bread and is fully grown, while in the cell of the youngest larva the original size of the cake can still be seen. The parti-

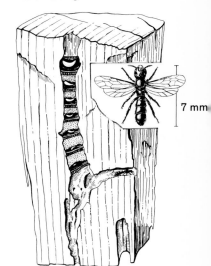

Fig. 27. Nest structure of the solitary bee Heriades. *The mother rests in the flight hole which, given time, she would also have closed with resin. An enlarged view of the bee to the right.*

7 mm

tions between the individual cells are made from resin. The mother bee herself is shown sitting in the flight hole which she would have sealed with resin had she not at that very moment been caught by an entomologist who added her together with her nest and her brood to his collection.

The greatest danger that threatens the brood of solitary bees, especially the brood of soil-inhabiting species, is the formation of mold. The methods employed to guard against this are many and varied. The plasterer bees, as we have seen, line the interior walls of their subterranean chambers with a waterproof wallpaper; others achieve a similar result by impregnating the cell walls with a glandular secretion or coating them with a waxy material. Some species of the genus *Halictus* (mining bees), which build their nests into clayey soils, construct, in addition, a ventilation system to insure low humidity inside the nest. After excavating a group of contiguous cells and impregnating their interior walls with a hardening secretion, they carefully dig away the clay behind this group of cells until they meet the hardened end walls from the other side. In this way, they produce a delicate comb of clay not so much by building it up but by cutting away what is not wanted. They carefully leave some slender supports of clay all around it (fig. 28). Two passages leading into the open, one up and one down, provide ventilation.

The most interesting nests are probably those made by the mason bees. This group includes the leaf-cutter bees (*Megachile,* or "Big-jaw"), the tailoresses mentioned

Fig. 28. Comb of the mining bee (Halictus quadricinctus) *in a clay wall exposed and partly cut open. Seen from above and from the side.*

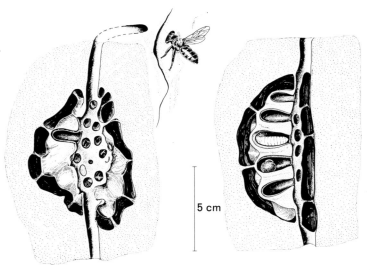

5 cm

above. They are about the same size as honeybees, some slightly bigger, others smaller. Telltale traces on the leaves of roses, lilac, raspberries, and other plants indicate where a leaf-cutter bee has used her sharp mandibles to cut out elliptical pieces from the edge of such leaves. She carries each piece rolled up under her body to her nesting site, which may be an insect hole in a piece of decayed wood, a crevice between two boards, or a cavity under a flat stone on the ground. She twists a number of such oval pieces into a thimble-shaped breeding cell, which she fills with a cake of pollen and nectar, and on top of which she lays her egg. Next she fetches more bits cut from leaves, circular this time, to make a lid. Dozens of such leaf thimbles, each with its lid made of several round disks, are strung together to form a cylinder somewhat resembling a cigar. In this case, the leaves serve as an insulating layer to protect the inside of the chambers from any dampness in the environment. Finally· the bee

a

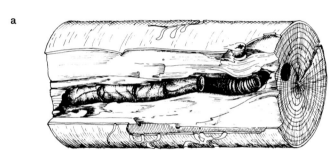

b

plugs the opening to the outside world by joining together a great many round disks like a stack of plates (fig. 29 a and b). If the flight hole happens to be not absolutely circular, the plug will not fit properly, and the way the leaf disks are pushed into the hole will sometimes look quite pointless. This is yet another instance of inherited behavior unalterably adapted to a standard situation and the inability to modify it by insight to any deviation from the standard.

The mason bee *Chalicodoma* prefers stones and rocks as sites for building her cells. From sand or weathered rock dust moistened with saliva she forms small oblong pellets which she carries, one by one, to her nesting site. There she builds cells from the rapidly hardening substance (fig. 30 and pl. 40 a and b, p. 78) usually in groups up to twelve. Liquid honey is stored as food. Into each cell the bee deposits a single egg, after which she walls

Fig. 29 a. A nest of a leaf-cutter bee (Megachile willoughbiella) *in a piece of wood, partly exposed. The flight hole can be seen on the cross section to the right.*
b. Rose leaf with two typical shapes cut out: ovals for the nests themselves, circles for the lids and for the plug in the entrance hole.

it up. Any remaining gaps are filled with mortar in such a way that the whole structure blends completely into the rock and becomes virtually invisible to the casual eye. Few would suspect the presence of larvae and their sweet provender under the smooth surface—they may grow up and pupate in complete safety. It is true that a hard task, literally, awaits them as they emerge from the pupal stage, for their strong jaws must gnaw them a way to freedom through a layer of concrete.

An amazingly complex system of protective measures is resorted to by another mason bee (*Osmia bicolor*). Her head and thorax are a deep black and her abdomen a vivid russet red (hence the name "bicolor"). First, she has to find an empty snail shell for each of her future offspring. In the depth of this she deposits some bee-bread, lays her egg on it, and closes the passage with a partition made of chewed-up leaves, taking care to leave

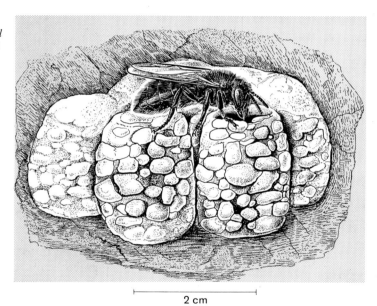

Fig. 30. The mason bee Chalicodoma muraria *on a completed cell not yet closed. The others contain a store of honey and an egg each, and are closed. The cells at the back have already been covered with mortar.*

2 cm

enough air space for the growing larva. Next, she fills nearly all the remaining whorls with small pebbles which she prevents from falling out by making another wall of chewed leaf pulp (fig. 31, right, below). Finally, she makes a series of flights to collect all kinds of dry stalks, blades of grass, thin twiglets, or even pine needles where available. From this material, she builds a tent-shaped roof over the snail shell (fig. 31, left), which eventually

hides it completely. She then proceeds to repeat the whole process for her next egg.

Hidden in grass or some small depression, naturally these nests are not easy to find. I shall never forget how I found my first. Once when I was out hunting insects, I made a swipe at an oddly flying, bizarre-looking creature. To my surprise, I found nothing in my net apart from a black-and-red mason bee and a dry stalk. Having read about the tent-building habits of these bees, I was intrigued. I released the bee and settled down to watch. After a while, I was rewarded by seeing her return, riding on another piece of stalk like a witch on a broomstick. The heavy load slowed her down to such an extent that I just managed to keep her in sight as I ran, and she herself guided me to her nest, which was hardly fifty meters away. Another time, I watched the same species at nest-building on a cow track in a mountain pasture. Here the stalks and twigs were all arranged toward the toe-end of the cows' footprints, a beautiful adaptation to local conditions. The speed of building was amazing. One of the bees I watched was just making a partition for which she had collected bits of leaves from a nearby strawberry patch. She kept flying back and forth in a straight line between the plants and the nest. Apparently other building materials, such as pebbles or thatching, are not picked up at random either, but are collected in quick succession from suitable preselected sites.

It is not difficult to understand how people become passionately interested in the fascinating behavior of soli-

Fig. 31. Nest of the mason bee Osmia bicolor. *The uppermost whorls of the snail shell contain the bee-bread with the oblong egg of the bee. The bottom whorl is blocked with small pebbles behind a supporting wall of chewed leaves. The snail shell is hidden under a roof of straws and dry twiglets.*

tary bees. But their less pleasing features must also be mentioned. Like all animals, solitary bees have parasites, and many of these come from their own ranks! More than a third of all known species of solitary bees are "cuckoo bees," which save themselves the trouble of building by smuggling their eggs into the nests of honest workers in an unguarded moment. The larvae of the parasitic bees grow faster than those of the rightful inhabitants and grab all the available food. Most parasitic bees specialize in particular kinds of "host" bees and are familiar with their nests and habits.

For certain species, however, the designation "solitary bees" is a misnomer. In some mining bees (*Halictus*), traces of a communal life may be found which suggest how, phylogenetically speaking, the evolution of bee societies may have come about. In the species *Halictus quadricinctus* (four-banded mining bee), the original builder of the clay comb (cf. fig. 28, p. 68) remains alive for a comparatively long time. She guards her nest and her brood and is still living when her offspring emerge. In some closely related species, mothers and daughters work together, thereby forming small colonies. The adults emerging from the first brood may be females with underdeveloped ovaries and act as workers. Fully developed females and males do not appear before the autumn. After mating, only the females hibernate and start their own individual colonies the following spring. Through this sequence of events, their social organization has actually reached that of a colony of bumblebees. In *Evylaeus marginatus,* another species of solitary bees closely related to *Halictus,* the founder of the nest may reach an age of five to six years and become the queen of a colony of over a hundred bees practicing division of labor. In fact, the transitional stages from solitary to a social life have been preserved more fully with the bees than with the wasps.

Bumblebees are better known and easier to observe than solitary bees. They shall serve us as examples for the building styles of social, yet primitive, members of the bee family.

HOMES OF THE SOCIAL INSECTS

I have already described one such community when I discussed the social wasps (see p. 56 f.). In this I may have been a little unsystematic, but the world of living things knows no sharp boundaries, and I hope therefore I may be forgiven for an occasional slight deviation from an ordered plan. However, I promise I shall be more methodical from now on, and shall discuss all other social insects in this chapter.

Bumblebees' nests

Despite their different appearance, bumblebees resemble solitary bees and honeybees so much in their anatomy and mode of life that they have been included in one family, that of the Apidae.

In early spring one can often see large bumblebees flying slowly along the ground as if searching for something. They are obviously not interested in flowers, though at other times they are busy enough gathering nectar and pollen. At this time of the year all the bumblebees we see are queens that have hibernated in some sheltered spot and are now looking for a suitable place to found a new colony. Some choose a site in the ground, such as an empty mouse hole; other species build their nests above ground on sheltered spots in a meadow, or on moss cushions near the edge of a wood, or even under a wooden floor. When the queen chooses her site and starts her nest, she is entirely on her own, exactly like a solitary bee. The sequence of events is essentially the same for all bumblebee species. We choose as our example the common field bumblebee of Europe, the carder bee (*Bombus agrorum*).

First, the earth of the chosen site is made level, cleaned, and covered with a layer of wax as a protection against damp from below. Wax is a fatty substance produced by the bumblebees in abdominal skin glands and secreted between their ventral and dorsal segments in the form of small flakes. They knead it with pollen and use the mixture to construct their breeding cells. The queen starts by building a single cell which she provisions with pollen; she lays half-a-dozen eggs into it and closes it with wax. Adjacent to that cell, she puts a rounded storage jar which she fills with honey as a reserve against rainy days. Then she surrounds the whole nest with a cover made of moss skillfully plaited together (hence "carder" bees), but leaves a flight hole to serve as a communication with the world outside. After hatching, the larvae feed on the store of pollen provided. Their mother opens the cell from time to time with her mandibles to replenish this, then she seals it up again. Other species of bumblebees fashion pockets for pollen on the inner wall of the cell and provide adequate amounts of food from the start. As the larvae grow, the wall of the cell bulges outward (fig. 32). When fully developed, each larva spins a cocoon in which she pupates. The queen, a thrifty manager, breaks down the remainder of the waxen wall with her mandibles to re-use the material. Some four weeks after the completion of the nest, the first batch of bumblebees crawl out of their cocoons. Having had to share their cell with their

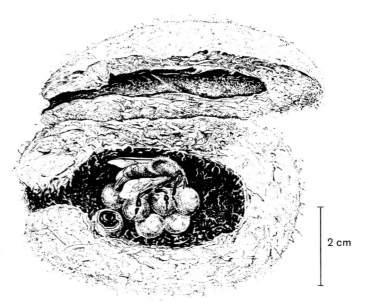

Fig. 32. A young nest of a common carder bumblebee, seen from above. The moss cover is folded back. Beyond the flight channel, the honey pot can be seen, bottom left. The queen is still alone. The growing larvae have caused the walls of the breeding cell to bulge.

2 cm

sisters, they are much smaller than their portly mother. To think that they are small because they are young would be completely mistaken. A winged insect having completed its larval and pupal phase is always fully grown. Our diminutive bumblebees remain small for the rest of their lives, which may last for several weeks. They are stunted because of the shortage of food and space they suffered during their larval existence, and they do not lay eggs. This remains exclusively the task of the queen, their mother. However, they help her by carrying in nectar and pollen, by defending the nest, and by building new breeding cells. The empty cocoons are not used again for oviposition though they are used as storage jars. The workers also enlarge the moss cover to gain room for the growing comb. Thanks to the hard work of these female helpmates, the next brood of larvae is better supplied with food and living space, and it develops into larger adults. Each brood gets progressively larger until finally, in summer, the queens for the coming year and the first males emerge from the cocoons. The various sizes shown in plate 42 (p. 78) were all found at the same time (September 2, 1935) in one nest of common European carder bumblebees.

In most bumblebee nests the comb never gets much bigger than the palm of a man's hand. Its cells, bulging and irregular, are partly built on empty cocoons. The whole can hardly be said to be a masterpiece of architecture.

The outer cover of the nest is more elaborate and varied. At my country place at Brunnwinkl in the Austrian lake district, a carder bee once built her nest in a shed where some roe deerskins had been hung up to dry. The queen had chosen a sheltered nest site under a box standing in the shed and had surrounded her nest with a warm and cozy cover made entirely from innumerable hairs of roe deer. When a bird's nest is chosen as a nesting site by the queen, as often happens, the community has less work to do. I have in my possession a wagtail's nest from a sheltered spot under the eaves of a boathouse. In its softly padded hollow I found, not eggs, but the nest of a bumblebee.

An even more cozy abode was chosen by a colony of early bumblebees (*Bombus pratorum*) that made their nest in a basket of chicken feathers. Unfortunately for them, this basket stood in a shed in Brunnwinkl, and its owner drew my attention to the bumblebees flying in and out of the shed. Soon the nest with it feathery cover was incorporated in my collection. When I opened it, I found the feathers immediately surrounding the hollow place that contained the comb stuck together to form a thick crust which thus made an excellent insulating layer between the nest cavity proper and the fluffy mass of feathers in which it was embedded (pl. 43, p. 79). I assumed that the material used for the purpose was either wax or resin collected from trees. It did not occur to me that it could be anything else, though I wondered how the bees had managed to work such hard substances into the loose mass of feathers. When thirty years later I wanted to know whether it was wax or resin, I was in for a big surprise. A chemical analysis found no resin and only traces of wax, which could be easily explained as impurities either adhering to the feathers themselves or accidentally introduced by the bumblebees. The crust itself dissolved in water. Its chief constituent was sugar! The bumblebees had obviously used either nectar or thickened honey from their storage jars to moisten the feathers of the inside boundary in such a manner that the whole dried to a dense, solid crust. I do not know whether this use of sugar as a form of cement has been observed before in bumblebees. It is unlikely that our early bumblebees use this building material at all regularly, for quite frequently they build their nests in the open, and a roof made of sugar would hardly make a good shelter against rain. But sugar was an excellent material for binding loose feathers indoors. Under what other conditions (if any) do these bumblebees use their sweet food for mortar? Have other bumblebee species made the same invention? There is still plenty of scope here for new discoveries by keen students of insects.

That some species of bumblebees fashion a nest cover from wax has long been known. The stone bumblebee (*Bombus lapidarius*), a species common in many parts of Europe, frequently builds subterranean nests in mouse holes or similar cavities and protects her comb against damp and cold by constructing such a wax cover (fig. 33 and pl. 44, p. 79). However, under more favorable environmental conditions one may find her nests lying in a cavity uncovered.

Yet, however well bumblebees choose their nesting sites, their nests cannot offer sufficient protection against the low winter temperatures of temperate latitudes. Nor will their accumulated stores, excellent as they are as reserves against spells of bad weather, take them through

Fig. 33. Subterranean nest of the stone bumblebee (Bombus lapidarius). *Part of the wax cover has been removed to show the comb with the queen. Several honey pots.*

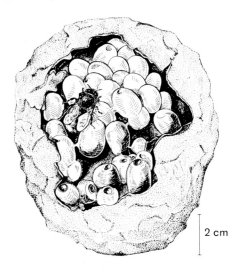

2 cm

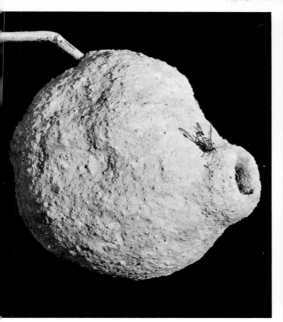

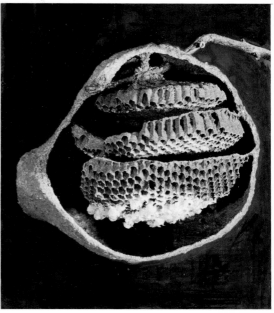

Plate 37 a. A nest of Polybia emaciata, *made of a mortar of clay and sand, seen from outside. One of the wasps sits in the flight hole, another above it.*

Plate 37 b. Combs with outer cover removed. The white caps on the lowest comb were lids of cells spun by larvae before pupating; shown here after lids had been gnawed through by emerging wasps. Diameter of nest is about 7 cm. (See pp. 62, 64.)

Plate 38. *Paper nest of wasps of the genus* Chartergus *(a South American genus, closely related to Polybia). The nest, which is 25 cm. long, has been cut open and the two halves are shown side by side. Each comb is attached to the outer cover and has a hole in the middle. The central shaft provides communication between the stories of this skyscraper and ends at the flight hole. (See p. 64.)*

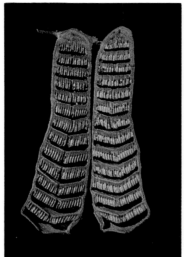

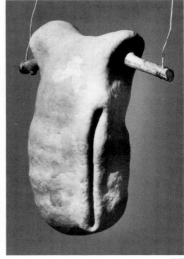

Plate 39. *The ceramic nest of Polybia singularis made throughout from a very fine clay. It is 30 cm. long, 15 cm. wide, and weighs 1350 grams. The long flight slit is seen on one side. (See p. 64.)*

Plate 40 a. *A female mason bee carries a small pebble to her nest in her jaws.*

Plate 40 b. *A half-finished nest with two cells still open seen from above. Near Rovinj, Yugoslavia. (See p. 69.)*

Plate 41. *Honeybees. Right: queen; center: worker; left: drone. (See p. 83.)*

Plate 42. *Common European carder bees, taken from the same nest on September 2nd. Left, the old queen; top and right, seven different sizes of workers; bottom, two young fully developed females, queens of the coming year. (See p. 75.)*

Plate 43. Nest of the bumblebee Bombus pratorum *in a basket of chicken feathers. Some feathers have been removed to reveal the nest. The innermost layer, which is stuck together, has been cut open and folded back. The layer of feathers is coated on the inside with a crust consisting essentially of sugar. (See p. 75.)*

Plate 44. The comb of stone bumblebees. Brown: larval cells; light color: cells covered with a spun membrane containing pupae. Near the edge, a group of honey pots. (See p. 76.)

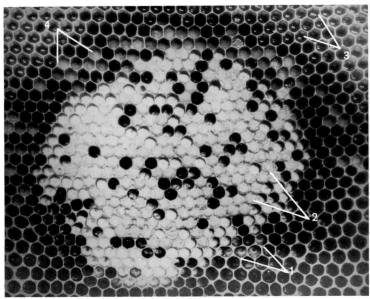

Plate 45. Brood nest of honeybees showing central part of the comb. (1) Open cells·with larvae; (2) lidded cells, holding pupae; (3) honey; (4) pollen. (See pp. 85, 93.)

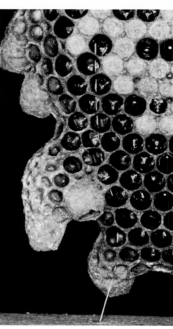

Plate 46. *The kind of home beekeepers offer their bees. The lid of the hive is removed, showing frames inside. One of them lifted out. The flight hole is in front, above the flight board. (See p. 82.)*

Plate 47. *A group of larger cells of a different shape. A limited number of these are built by the workers for the raising of queens. (See pp. 84, 85, 89.)*

Plate 48. *A comb in process of construction. It has been started in three different places, but the originally separate tapering sections are already touching each other at top. Note perfect fit of cells where two sections join. This comb contains mainly cells for the raising of male larvae (drones). Smaller cells designed for larvae of workers are seen in the upper part. Parts that are light in tone are made of new wax. Darker parts indicate reused wax. (See pp. 85, 88.)*

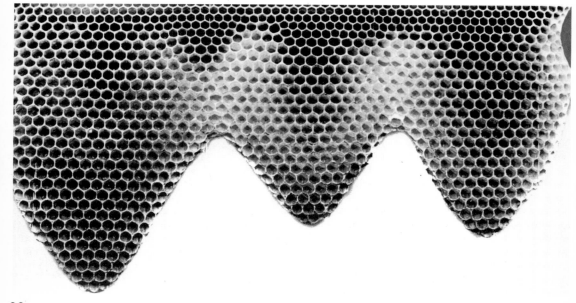

a whole winter as in the case of the honeybees. The colonies all die in late autumn, and only the fertilized queens, having sought suitable shelter before the onset of bad weather, survive in a state of torpor until the next spring.

Life expectancy is different in hot countries where colonies of tropical species may live for several years. It is different again in the cold north, where bumblebees are also found (for the family has spread successfully over most of the globe). Conditions of life in, say, Lapland or Greenland, are too harsh for the development of successive broods of auxiliary females, and the bumblebee has to live the life of a typical solitary bee. She alone carries the whole work load during the short arctic summer, but during that brief period of abundant flowering she feeds her larvae so well that her very first brood reaches sexual maturity and ensures the survival of the species from one year to the next.

Honeybees and their dwellings

All the insects we have discussed so far build their own homes. The honeybees, however, live in dwellings put at their disposal by the beekeeper. They have only to supply their own furnishings and fittings, which consist of the waxen structures of their combs. Obviously, this has not always been so, for bees lived on our planet millions of years before the first humans appeared. Yet even before man decided to take them under his care, bees did not actually construct their dwellings. They made their nests in hollow trees, or, where trees did not grow, in holes in the ground, or in crevices of rocks. In such hidden places their stores of honey and their brood were well protected, especially as the bees themselves were armed with venomous stings to defend them. However, not *all* enemies could be kept away. Bears, protected against stings by their thick fur, have always searched in hollow trees for bees' nests to rob them of their honey. Man may actually have learned from bears how to locate and plunder the nests of wild bees. Later, man thought of a better method: he provided the bees with artificial dwellings close to his own home so that he could tend them and guard them like domestic animals. Man became a beekeeper.

Homes that beekeepers provide and bees fit out. The history of beekeeping goes back into the distant past. We know from illustrations on Egyptian temples and in royal tombs that five thousand years ago the Egyptians kept their bees in clay pipes made from the mud of the Nile. Such pipes are still used by them today. The dwellings that men offered to their honey suppliers have varied greatly. Where wood was plentiful, as in wooded parts of Europe, pieces sawn from a hollow tree or logs specially hollowed out for the purpose were often used. Such habitations closely resembled the natural homes of the bees. In some areas bees were kept in wicker baskets. However, wooden hives were the most widespread. In the nineteenth century, movable frames were put into hives for the bees to build their combs in (pl. 46, p. 80). They were an important invention because these "movable combs," which can be removed one at a time, enable the beekeeper to appropriate surplus honey without disturbing his bees or destroying them. In the past this could not always be avoided.

These are only a few examples of the type of homes beekeepers have invented for their bees. The bees themselves and their combs have never changed. They have not become domestic animals in the true sense of the word. They are not like dogs, for example, which have become the companions and willing helpers of man and have, by breeding, developed the greatly differing physical and psychological characteristics of dachshunds, sheep dogs, huskies, and so forth. Bees will accept a hive that is offered them because it suits their requirements. But their manner of constructing honeycombs, their other interior arrangements, and, indeed, their entire behavior is based on primeval, deep-rooted instincts that men are unable to influence. The differences in evolutionary history and organization that separate us from the insects are immeasurably greater, of course, than those which separate us from vertebrate animals. We can read in the eyes and movements of a dog the expression of psychological states akin to our own. There is no intuitive access to the inner life of bees.

Their achievements, however, not the least of which is their amazing work as builders, have long been the admiration and wonder of mankind. To appreciate them we need to know the rudiments of the structure and organization of a bee community.

The bee community. Unless the beekeeper increases their numbers intentionally, a colony of bees consists of some forty thousand to eighty thousand individuals, corresponding in numbers of inhabitants to a modest town. All the bees are the offspring of the queen, the only egg-laying female in the colony (pl. 41, right, p. 78). The mass of the population consists of worker bees (pl. 41, center), also females but with poorly developed ovaries which, under normal circumstances, do not lay eggs. But they should not be confused with "auxiliary females" like the workers of bumblebees and wasps; the worker bees differ from the queen not merely by the absence of properly developed ovaries, but by the possession of special organs for certain functions which are either lacking or less well developed in the queen. Their hind legs are equipped with brushes and baskets for the collection and transport of pollen (pl. 50, p. 97). Their tongues are longer than the tongue of the queen to enable them to reach the nectar at the bottom of deep-cupped flowers. They possess glands on the ventral side of their abdomen that secrete wax for the building of combs (fig. 37, p. 88). By these and other physical characteristics, the workers are clearly distinguished from the queen; they form a separate "caste." In spring and summer the colony contains males, or drones, which are bigger and less agile than the workers (pl. 41, left, p. 78). They have no function other than that of mating with the queen on her nuptial flight. Once this is over, they are stung to death by the worker bees in the famous "battle of the drones," or are driven away and left to die of hunger.

Unlike the queens of bumblebees and wasps, a queen honeybee never founds a colony on her own. She never helps in the building of cells, nor gathers food, nor tends the brood. All this is the exclusive domain of the workers which practice a system of division of labor. However, they do not specialize in a particular job, such as building, doing nothing else. Their occupations change with age. By the time they reach the end of their lives, that is, after about five weeks in the case of bees that emerge in spring, they probably have performed every kind of work that has to be done in a hive—or might have been. During the first ten days of their lives, the young bees are occupied with domestic duties, especially with the tending and feeding of the brood. They not only feed the larva honey, but a kind of bees' milk called "brood feed" as well. This is a nourishing juice produced by modified salivary glands. As these are fully developed only in young worker bees, the care of the brood is an occupation of a bee's early life. In a second phase of development, the glands producing the brood feed dry up and the tending of the brood comes to an end, while the wax glands now reach full functional maturity and the workers take up the job of building. At the same time, they attend to various other duties, mostly inside the hive. In a third and final stage, the wax glands cease to function and the bees become outdoor workers, gathering nectar and pollen. Under favorable conditions, these new tasks are performed with great zest and with a highly efficient organization of labor. Some individuals fly out to reconnoiter for flowers rich in nectar or pollen, and, by dancing on the combs, inform their mates at what distance and in what direction these sources of food are to be found. When indicating a rich source of nectar and pollen, they dance with greater intensity and persuasion than for a poorer one. In this way, the size of the groups that respond is always related to the potential yield of the source. Since bees are most exposed to danger when working outside the hive, it is with good reason that food-gathering is the last of their many occupations.

Worker bees emerging in late summer and autumn live for several months; being well fed and less worn out by work they can survive the winter. A queen can carry out her maternal duties for four or five years. Should she die or fail to function, a new queen may be bred from a young larva. The life of a bee colony need never end unless it succumbs to disease or accident. However, this does happen quite often, and to safeguard the survival of the species, bee communities must multiply. They do this by swarming. One part of the population of the hive moves away and founds a new colony. To make this possible, a new queen is required.

In springtime, when abundance of food swells the numbers so that living room becomes scarce in the hive, the workers start building a few cells that are larger than the rest (pl. 47, p. 80), and by special feeding raise queens in them. Several queens are raised at one time, to be on the safe side. A single queen might meet with an accident, and, not infrequently, a colony produces more than one swarm in a summer. Shortly before the first of the new queens emerges from her cell, the old queen leaves the hive in a frenzied flight, accompanied by half the community. Soon the swarming bees settle around her in a large cluster—usually on some nearby branch where an observant beekeeper can gather them in and transfer them to an empty hive. If he fails to do so quickly, the swarm will be lost to him. Scout bees will have been busy searching, and as soon as they discover a suitable nesting site, they inform the waiting swarm in bees' language—by dancing on its surface—in what direction and at what distance the new home is to be found: it may be a hollow tree or, perhaps, an empty hive belonging to another beekeeper some distance away, in which case the rightful owner is the loser. For the cluster now dissolves and the swarm makes a "beeline" for its new abode. The bees take enough food for a few days by filling their honey stomachs from the stores before leaving the old hive. But they cannot bring any combs along, so fresh ones have to be constructed with great haste. As all age groups are represented in the swarm, there will be no shortage of building labor. The urge to build is particularly strong in swarming bees; moreover, their wax glands develop at an earlier age and stay functional longer than at other times. Actually, the division of labor into age groups is never very rigid. The sequence of activities in time is, in fact, more flexible than was formerly believed.

Even without swarming, there is always plenty of work for building bees. When the harvest is abundant, the old combs have to be enlarged and new combs built. When the season comes round for the raising of drones and young queens, larger cells for their larvae have to be constructed, and in addition there are always all sorts of small jobs to be done about the hive.

Construction of the comb. It is important to remember that in contrast to the paper combs of wasps, which normally are horizontal and have cells only on their undersides (cf. p. 61), the combs of bees are made of wax, hang vertically, and have cells on both sides separated by a wall in the middle (fig. 34, p. 86). They serve for the rearing of the brood and for the storage of honey and pollen. The same kind of cell is used for all these purposes. Within the building precinct of a hive, the queen bee lays her eggs in the central areas of the combs; only the more peripheral parts are used as containers for pollen and honey (pl. 45, p. 79). The larger cells for the rearing of drones are usually placed in the lower sections of the combs (pl. 48, p. 80). Even larger cells, shaped somewhat like spruce cones, which are designed for the raising of queens, are usually added a little to one side or at the lower edge of the main cell matrix (pl. 47, p. 80).

The walls of the main body of connected cells form regular hexagonal prisms (cf. fig. 34). This is a note-

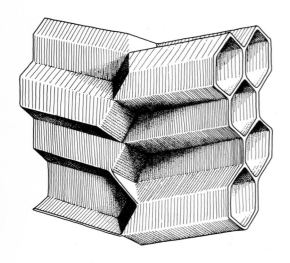 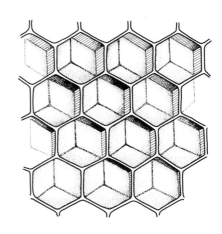

*Fig. 34. Detail of a honey comb.
Left, side view. A few cell walls
have been removed to reveal parts
of the bottoms of some of the cells.
Each bottom is fashioned from
three flakes of wax; it is shaped
like a shallow concave triangular
pyramid and constructed from
three equilateral rhombs. The cell
walls are slightly tilted toward the
rim, which prevents honey from
running out before cells are
closed. Right, cells viewed from
the surface of the comb. Drawn
after a model by M. Renner.*

worthy fact. After all, the bees might build their cells with
rounded walls as the bumblebees do or as they themselves
build for the cradles of their queens. Or they could base
their architectural style on some other geometrical con-
figuration. However, if the cells were round or, say, oc-
tagonal or pentagonal (fig. 35, top), there would be
empty spaces between them. This would not only mean a
poor utilization of space; it would also compel the bees to
build separate walls for all or part of each cell, and entail
a great waste of material. These difficulties are avoided
by the use of triangles, squares, and hexagons (fig. 35,
bottom). The three-, four-, and six-sided shapes of figure
35 are carefully drawn to enclose the same area. Provided
their depth was the same, such cells would therefore hold
the same volume. But of the three geometrical figures
equal in area, the hexagon has the smallest circumference.
This means, of course, that the amount of building material required for cells of the same capacity is the least
in the hexagonal construction, and hence that such a pattern is the most economical design for warehouses.

In most combs, one edge of the hexagonal prism points
to the top, and another to the ground. Exceptionally,
combs are found in which there are horizontal walls at
the top and bottom of the cells (fig. 36). The question
whether the customary arrangement has a static advantage that increases the load-bearing capacity of the cells
has never been investigated, as far as I know. Dr. Georg
Kirchner of Frankfurt am Main very kindly agreed to

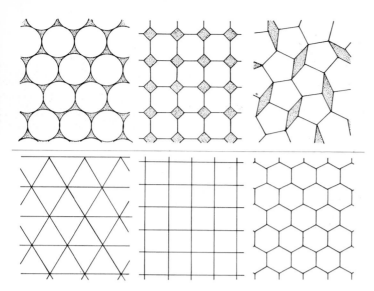

Fig. 35. *The advantages of the hexagonal shape. Explanation in text.*

Fig. 36. *Above, normal cell construction. Below, exceptional orientation of cells.*

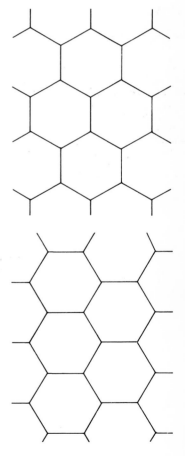

look into it for me as an engineering problem. His computations gave the same result for both arrangements and do not help us to explain why the bees prefer one to the other. The geometry of the shape and depth of the cell bottoms and the manner in which they dovetail into each other contributes, no doubt, a great deal to the stability of the comb (fig. 34). Anyone lifting a full honeycomb for the first time will find it amazingly heavy. A comb measuring 37 by 22.5 centimeters can hold more than four pounds of honey. Yet in the manufacture of such a comb, the bees use only about forty grams of wax. The relationship between the construction of a comb and its strength would seem to be a worthwhile subject for study.

Building can proceed quite rapidly if necessary. When bees start building, they first attach themselves to each other in chains. Soon they form themselves into a dense ball, the building cluster within which they maintain a temperature of 35° C. (95° F.)—the temperature needed for the secretion of wax. In honeybees, wax is formed only in glands on the underside of four abdominal segments. Where these segments meet, wax appears in the shape of two small flakes between the ventral scales (fig. 37, left). To pick up a flake, the bee uses the greatly enlarged first tarsal joint of her hind leg, which carries on its inner side a small brush for the collection of pollen. She impales the little flake on the last row of bristles (fig. 37, bottom right), pulls it out of its skin pocket, and passes it forward to be taken over by her front legs and

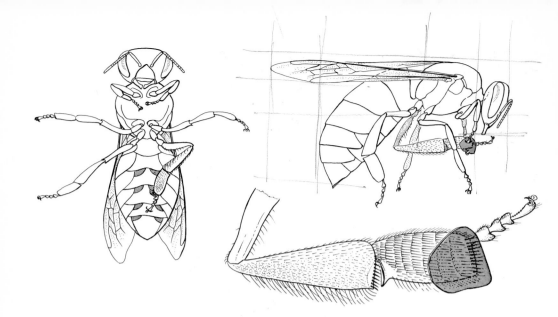

Fig. 37. The drawing shows how a bee pulls the wax flakes she has secreted from her abdominal pockets and passes them to her mouth for "processing." Left: bee seen from ventral side; wax flakes can be seen protruding from the glandular pockets. The bee is about to pick up one such flake with her left hind leg, transfixing it with the bristles of its enlarged first tarsal joint (left and bottom right). The flake is transferred from the hind leg to the front legs and mandibles (top right).

mandibles (fig. 37, top right). Holding it in front of her mouth, the bee kneads and mixes it thoroughly with a secretion of her salivary gland. This treatment, combined with the right temperature, gives it the necessary homogeneous consistency and that degree of plasticity at which it can best be molded.

Building starts at the ceiling, which in a normal hive means at the top of a frame. Bees usually begin at two or three different places at once and construct sections that taper downward. They do not build one complete cell after another. While the lateral walls of the first cells are gradually being added to, new adjoining cells are being started lower down. As these triangular sections are enlarged laterally, they gradually coalesce from the top down. The joins are so skillfully made that no traces of the separate beginnings remain visible (pl. 48, p. 80). This is even more remarkable when one considers that many bees are employed in the building of each individual cell and that they often relieve each other at intervals of no more than half a minute or so. Apparently, each bee immediately comprehends what stage the construction has reached at the place where she starts to work and continues accordingly. When the pace is fast, the individual sections grow rapidly, both vertically and laterally, so that a contiguous building front is reached in a short time.

Modern beekeepers insert artificial walls with a stamped-on hexagonal pattern into the frames they want their bees to fill with combs. This speeds up the building program, partly, of course, because the bees need less wax. However, the cells built without such aids are no less regular. It has often been maintained that this regularity was not a particularly remarkable achievement since, under the influence of lateral pressure, the cells would of necessity acquire the shape with the smallest surface area. This is not so. From the very beginning, the cells are built in this, the most practical, shape. Usually a rhomb-shaped section of the base is made first, followed by the first parts of two adjoining cell walls. Next, another rhomb is added and two more cell walls are started; thereafter, the third rhombic section is erected, which completes the hexagon, and the two last walls. Right from the start, the cell walls meet at the correct angle of 120°. Admittedly, the regular hexagonal shape is hidden at the top by a ring of wax which contains material for further use, but as soon as this bulky wax ring is carefully removed, it shows at once. This pattern must be imprinted into the minds of the bees by the forces of heredity. The bees even apply it as useless ornamentation to the outer surfaces of queen cells in the course of their modeling work (cf. pl. 47, p. 80). The same regular hexagonal shape characterizes the combs of the paper wasps and those of the tropical wasp species that are modeled in clay (pl. 37 b, p. 77).

It is not only the exact shape of the cells that depends upon the skill of their builders; skill is just as much needed to vary the size of the cells for worker bees and drones, to manufacture such extraordinarily thin walls, and to orient them accurately in space. None of these things just "happen," they are the result of work directed to a purpose.

The cell walls are built with a gradient of about 13° from base to opening (fig. 34, p. 86). This is sufficient to prevent the thick honey from running out. The distance from the cell wall to that opposite is 5.2 millimeters in a worker cell, and 6.2 millimeters in a drone cell. The thickness of the cell walls is 0.073 millimeter, with a tolerance of no more than 0.002 millimeter. What truly astounding precision! Economy in the use of building material is thus taken to the utmost limit. Human craftsmen could not do work of this nature without the use of carpenters' squares and sliding gauges.

Measuring instruments of the bees. The bee's own head serves as a plummet to determine the line of gravity. It

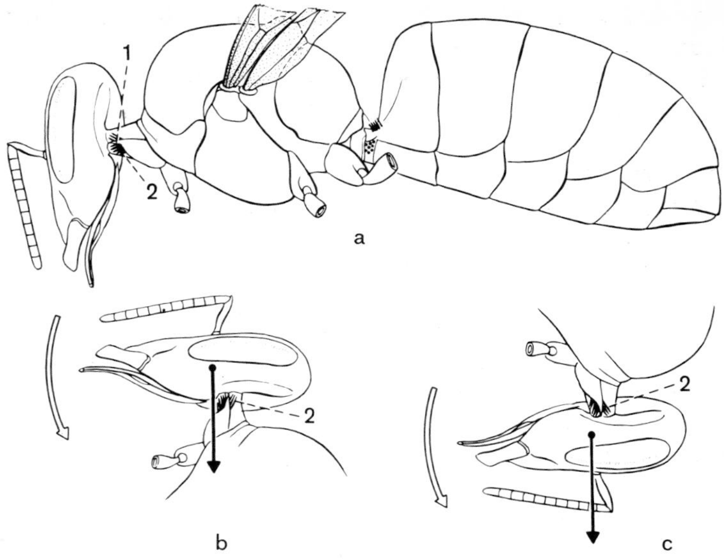

Fig. 38. Sense organs for perceiving the direction of gravity. The head of the bee rests on two pivots forming part of the front thorax (1). Its center of gravity is slightly lower than its point of suspension. Hence, the head is pulled toward the thorax if the bee's head is above the rest of her body (bottom, left) and toward the back if it is pointing downward. These rotary movements are minutely registered by the sensory bristles which touch the back of the neck (2).

rests on two pivots forming part of the outer skeleton of the thorax (1 in fig. 38 a), and its center of gravity lies below this articulated connection. Hence, if a bee sits with her head pointing upward, its heavier, lower part will be pulled toward the thorax by the force of gravity (fig. 38 b, light arrow). In a downward position, the head is automatically rotated in the opposite direction (fig. 38 c.). These gravity pulls are accurately registered by a tactile organ consisting of a set of highly sensitive bristles (2) on the tips of these pivots. Any position at an angle to the vertical is registered by a characteristic distribution of pressure on the set of sensory hairs. This is the way bees control both their own position in space and the position of the comb, which is always built vertically downward.

It has been possible to prove experimentally the importance of these sensory organs in the bees' necks for their building activities and for the correct orientation of the cell walls. These organs can be put out of action by coating the bristles with a warm mixture of wax and rosin which hardens on cooling and by which they are completely immobilized. Setting up this experiment was a time-consuming operation. It was necessary to guard against the possibility of results being affected by bees outside the usual age groups taking part in building. Therefore, all bees in the experimental colony—some thousand individuals—had to be treated. The effort proved worth while. The bees were in a building mood and immediately gathered in a building cluster. But nothing happened. They behaved like workmen whose tools have been confiscated: they did not work. The gumming-up of their neck bristles had clearly not impaired their general well-being, because the flying-out and food-gathering activities of the hive were normal. Nor was there a reduction in their production of wax, but the flakes were allowed to drop to the ground unused. In one experiment in which a building cluster was observed for a whole month, no more than three pitiful little cells with irregular walls were produced during the first two weeks (fig. 39: 1 and 2). Then came a heat wave. The coating substance started to melt, and the tips of the sensory hairs reappeared. The result was that the bees, within the next four days, after producing some irregular transitional shapes, managed once more to build more or less normal cells (fig. 39: 3–6). The experiment indicates that for the correct positioning of their cell complex in space, the bees need the organs in their necks.

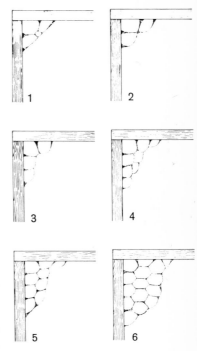

Fig. 39. *When their neck organs were temporarily put out of action, the bees, though exhibiting great eagerness to build, managed only three round cells over a period of fourteen days (1 and 2). When their neck organs began to function again, they achieved after the initial production of some irregular transitional forms (3–5), a number of correct hexagonal cells (6).*

The search for the instrument used by the bees to determine with such impressive accuracy the size of the cells and the angles and dimensions of the cell walls has not led to conclusive results so far. However, experiments on queens have produced some interesting evidence. A curious fact that has long been known about bees is the way in which sex is determined. A queen can either lay unfertilized eggs, which produce males (drones), or fertilized eggs, which produce females (queens or workers). She possesses a sperm pouch where the male sperm, acquired by her on her nuptial flight, is stored for the rest of her life. From this she can release sperm cells the moment an egg slides past its exit duct. As the larvae of the drones grow bigger than those of worker bees, the queen lays her unfertilized eggs in the larger drone cells. If the tips of her front legs are amputated, the queen continues her egg-laying unabated, but she can no longer distinguish between the two types of cells and produces a terrible muddle of fertilized and non-fertilized eggs. From this experiment, one may infer that the queen measures the size of the cells in which she is about to lay her eggs with the tips of her front legs; hence it is probable that the worker bees do likewise.

The problem of discovering how the building bees gauge the thickness of the cell walls seemed, if anything, even more difficult to solve. Obviously they must be able

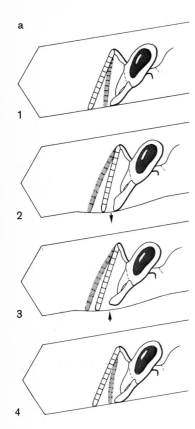

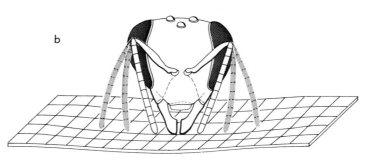

to measure, for how else could they keep to the exact thickness of 0.073 millimeters (0.094 millimeters for drone cells)? Yet this particular puzzle has been solved.

When the bees start building up a cell wall, they first make a roughly kneaded ring of wax at its upper end and gradually roll it out into a thin lamella by planing motions of the mandibles. Throughout this operation, they constantly measure the thickness of the wall and shave off surplus wax. From the way they do this, one might think that the workers had received a thorough training in physical mechanics. Only the basic principle involved can be explained here very briefly. The bee presses her mandibles against the cell wall and thereby produces an indentation (fig. 40 a: 1 and 2). When she withdraws her

Fig. 40. Checking the thickness of the cell wall by measuring the elastic resilience of an indentation with the tips of the antennae: (a) side view; (b) front view.

mandibles, the wall returns to its original shape (3, 4). During this process, she continually examines the relevant section of the wall with the tips of her antennae (fig. 40 b). These tips are endowed with special sense organs consisting of three rings of tactile cells equipped with curved hairs, each surrounding a cell sensitive to tactile and chemical stimuli and carrying a spike. This spike is pushed into the wall while the bristles of the rings are placed upon it. In this manner, the organ registers the course of deflection and recovery. Under the conditions prevailing—a given consistency of the wax, a constant temperature of 35° C. (95° F.), and a predetermined shape of the cell—the speed of this movement depends entirely on the thickness of the wall and reflects it accurately. When the tips of the antennae where these sense organs are located were experimentally removed, this abruptly put an end to the bee's precision work. The bee still managed the basic architecture of the cells, but the walls got either too thin, or, more often, too thick (pl. 49, p. 97; compare these irregular cells with the thin walls and regular edges of normal newly built cells shown in pl. 48, p. 80).

Storage cells for pollen are left open. Honey cells are closed with a wax lid when they are full. The wax necessary for this is kept ready for use as a thick ring around the lip of the cell, and can be quickly rolled over the opening from all sides when it is required. The breeding cells, too, are covered with a domed lid of wax for the twelve days of pupation (pl. 45, p. 79). Below the wax lid, the larvae themselves spin a dense cap of silk threads. Under this twofold cover, the metamorphosis of larva into winged bee can proceed undisturbed.

Orientation of combs by the earth's magnetic field. By putting wooden frames into his hives, the beekeeper determines the direction in which the combs will be built. But the interesting question is, what happens under natural conditions, in a hollow tree, for example? How is it possible that the thousands of bees, starting work without delay after taking possession and often completing large parts of their new combs overnight, end up, not with a chaotic muddle, but with a well-laid-out, regular structure?

In this context, the German ethologist Professor Martin Lindauer and his co-worker Dr. Martin Oehmke have made a remarkable discovery. They placed a swarm from a conventional hive into a cardboard cylinder that had no frames and whose flight hole was in the center of the floor. Neither the cylinder nor the position of the flight hole gave any pointer to the orientation of the combs in the old hive. Yet, within a few hours, the bees had produced the beginnings of neat, parallel combs whose orientation corresponded to that of the combs in the original colony, or deviated by no more than a few degrees (fig. 41). How was this possible? Recently it has been found that the orientation of bees is influenced by the magnetic field of the earth. Though the experimenters had carefully removed all directional landmarks from the bees in the round hives, they had not been able to take away the natural compass of the magnetic field. Their assumption that this was the bees' means of orientation was confirmed by further experiments. A swarm was moved first to one cylindrical hive and transferred from there into another, similar one. The combs were placed in exactly the same compass direction in both hives. Next, the natural magnetic field was artificially disturbed and deflected by 40°

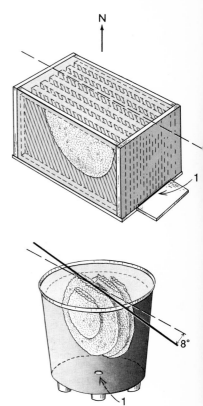

Fig. 41. Bees from a normal hive (top) were moved into a cylindrical container with a flight hole in the middle of its floor (bottom). The orientation of the combs corresponded almost exactly to that of the original hive (broken line). The bees had oriented themselves by reference to the earth's magnetic field. (1) Flight hole. The heavy arrow points due north. Diagram after M. Lindauer and M. Oehmke.

between the first and second transfer with the aid of a magnet placed outside the second hive; now the direction of the new combs differed from the previous direction by the same angle of 40°.

Before this discovery, it had occurred to neither bee-keepers nor scientists that bees laid out their combs with a compass. But now that they have demonstrated their ability to do so, it is easy to see what a useful accomplishment this is. For when a swarm has taken possession of a dark hollow in a tree, thousands of workers start building at the same time at different corners of the new abode, and there is no foreman or site architect to tell each worker what to do. However, if they all have the urge and the capacity to orient their combs in space exactly as in their old home, a well-ordered structure will be the result. The manner in which the bees perceive the magnetic field of the earth is still a mystery.

Bees' glue (propolis). Though wax is the chief building material of bees, it is not the only one. They also use resin for filling in gaps and holes. This same material also has an entirely different use. When a mouse or other small animal invades the hive in search of honey and is killed by the valiant defenders, its dead body is covered with a layer of resin which effectively shuts out the air and so preserves and mummifies it. Wax is produced by the bees themselves, while resin, called propolis, has to be collected. The bees gnaw it with their mandibles from the sticky buds of certain trees, transfer it to the "baskets" on their hind legs, and carry it home like pollen (pl. 51, p. 97). However, this material, which is viscous and sticky, cannot be transferred to the leg baskets in flight in the manner of pollen. The bees have to alight somewhere to do this. It is surprising that they can handle it at all, but the same secretion of their mandibular glands that helps to improve the working properties of wax is of use here too.

The behavior of a bee arriving home with her baskets full of propolis is quite different from that of a bee carrying in pollen. The pollen collector seeks an empty cell to unload her harvest. But a bee that carries propolis makes her way to the building site where her product is needed and very unobtrusively offers it to the building workers. Sometimes she may sit or slowly walk about for hours. When the building workers need propolis, they come along and gnaw off the required small amounts from the tough lump in her basket. At times propolis is mixed with wax to make it go further.

The chief time for collecting propolis comes in late summer or in autumn when the nights get cool and draughty places in the hive become menacingly noticeable. During that period the bees are hard at work filling in cracks and crevices with glue to preserve the precious heat of the hive. Sometimes the bees have been so busy that when spring returns the beekeeper can hardly separate the frames or prize them from the walls of the hive.

In southern countries, heat rather than cold is the danger since the wax structures will melt at high temperatures. Here again, the bees know how to cope with the problem. In the hot volcanic areas of Salerno in southern Italy, bees were observed mixing propolis into the wax they used for building in order to raise its melting point.

Manifold are the benefits and pleasures we owe to the bees. A less well-known reason for our gratitude is the fact that in all probability propolis was one of the secret ingredients that the Italian makers of violins added to the lacquer to improve the sound of their instruments.

Moving house. The honeybee (*Apis mellifica*) is found today in practically all countries of the world. Where it did not occur naturally, people sought to introduce it. because no one wanted to be without the help of these assiduous collectors of honey, producers of wax, and pollinators of flowers. Originally they came from India, the only country where more than one species of the genus *Apis,* or honeybee, exists today. The smallest and the most beautiful among them is the dwarf honeybee *Apis florea*. These bees are about the size of houseflies, brick-red with silvery-white, furry bands. Their nest consists of a single comb, the size of a hand, built in the open on the underside of a branch. On a study tour in Pakistan, Dr. N. Koeniger, a German scientist, observed a colony that had left its comb to build a new one at a different place—a common occurrence. The next day the worker bees returned to the old comb, gnawed off the wax of the old cells, and carried it in their leg baskets to their new abode (pl. 52, p. 97), thus saving themselves a great deal of energy in wax production by their thrifty action. The same behavior was observed by Professor Friedrich Ruttner at the Institute of Apiculture at Frankfurt University when a colony of dwarf honeybees moved from one corner of the observation room to another.

No doubt beekeepers will be interested in stingless bees. This subfamily, the Meliponinae, is widespread in tropic regions, especially those of the American continent, and contains many species. Maybe apiarists would have replaced stinging bees with stingless kinds a long time ago if these species were not less productive. Besides, though their sting has degenerated, they are no less formidable in defending their homes—they bite, and their bites are very painful.

They, too, use wax as a building material, but often mix it with other substances such as resin or soil particles. They employ it to build breeding cells, rounded storage jars for honey and pollen, and often also covers for their combs. In some species the nests are rather primitive, more like those of the bumblebees, but in other, highly organized South American species of *Melipona,* the nest structures are large and impressive. The swarming behavior of these communities differs from that of the honeybees in the sequence of events. In both groups young queens are raised beforehand. But the stingless bees start building a new home long before the new queen is ready to move in, and instead of producing their own wax, the members of the new colony dismantle part of their old home, and transport the material to the new site to use it once more. They also take honey and pollen from the larder of the mother hive to fill their own storage vessels in advance of their move, which is thus literally a move "lock, stock, and barrel." In honeybee colonies it is the old queen who leaves the hive during swarming and is replaced by a young successor. But the old *Melipona* queens are too fat and sluggish to travel, and the workers conduct a young queen to their new abode. Occasionally members of the newly established home go on with the removal of building material from the old nest while they are already raising their own brood. Thus, the process of swarming combined with that of moving house may take several weeks.

Homes of the ants

"Anthill" is the name we give to the large mounds of conifer needles, dry stalks, and broken twigs constructed by the red wood ants (*Formica rufa*) and to the small soil mounds of the small yellow meadow ant and the common brown garden ant (*Lasius flavus* and *Lasius niger*). These mounds, the most conspicuous and best-known structures erected by ants in Europe, appear to

Plate 49. Comb built by bees whose antennal tips had been amputated. The cells are irregular; their walls are excessively thick in some places, excessively thin in others, and there are holes. (See p. 92.)

Plate 50 (top right). Bee with baskets filled with pollen. (See p. 83.)

Plate 51. Bee with baskets filled with propolis. (See p. 94.)

Plate 52. Workers of the dwarf honeybee gnawing wax off an abandoned comb and putting it into their leg baskets. (See p. 95.)

Plate 53. Anthill of the small red wood ant (Formica polyctena). *(See p. 107.)*

Plate 54. Cross section through a nest of Formica polyctena: *(1) mound; (2) excavated sand; (3) soil nest; (4) tree trunk. (See p. 107.)*

Plates 55 a and b. Transverse and longitudinal sections through a larch stem inhabited by carpenter ants (Camponotus herculeanus). *(See p. 110.)*

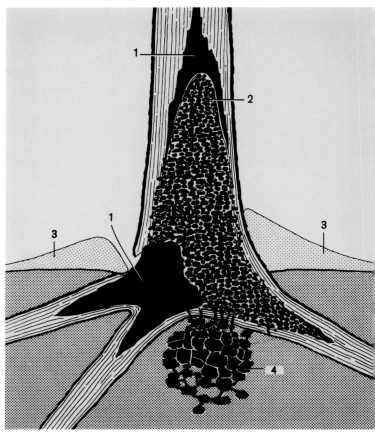

Plate 56. Carton nest of the jet ant (Lasius fuliginosus) *in a hollow tree, shown diagrammatically. (1) Tree cavity; (2) carton nest; (3) excavation; (4) winter nest. (See p. 110.)*

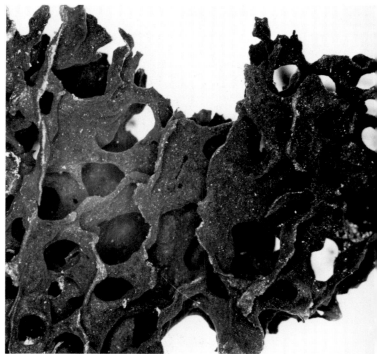

Plate 57. Carton nest: part of a nest of Lasius fuliginosus. *(See p. 110.)*

have absolutely nothing in common with the marvelous waxen structures of the bees. The wingless inhabitants that mill apparently aimlessly upon and around the nest do not appear closely related to the winged friends of the flowers that can be seen flying from their hives with such perfect sense of direction and purpose. Nevertheless, the ants (family Formicidae) and the bees (Apidae) are grouped together in the same order Hymenoptera because, in spite of all the obvious differences in appearance and behavior, the main features of their anatomy and social organization are very similar.

There are over six thousand known species of ants. Not one of them is solitary. All are social, though some colonies may have hardly more than a dozen members. Others may number thousands, hundreds of thousands, or even over a million individuals.

Castes of the ants and their tasks. As with the bees, the mass of the ant population consists of workers, females with undeveloped ovaries, unfit to propagate but capable of carrying out all the other tasks necessary for the continuance and well-being of the colony. Having no wings, they are earthbound, but the comparative slowness inherent in their pedestrian status is offset by their enormous numbers. The sexual ants are winged, but the females cast off their wings immediately after their nuptial flight, and the males die soon afterward. As with the bees, the population thus consists of males, females, and workers (fig. 42 a). But whereas all worker bees look alike, ant workers have differentiated into a number of "castes." In the wood ants, for instance, we find, apart from the queens, not only workers of medium size, forming the most numerous group, but also larger and smaller individuals in numbers that decrease as the sizes get more extreme. Size differentiation is to some extent related to the division of labor. Small individuals tend to work inside and large ones outside the nest. In other species, two or more castes of workers with sharply differing bodily characteristics may occur without any transitional forms. For instance, many species contain, in addition to the ordinary workers, ants with very large heads (fig. 42 b). They are often called "soldiers," but they are not necessarily more bellicose than the rest. Their main task may be the crushing of hard foods, such as hard seeds or insects with strong armor plating. For such a task, which can be carried out sitting peacefully inside their nest, the

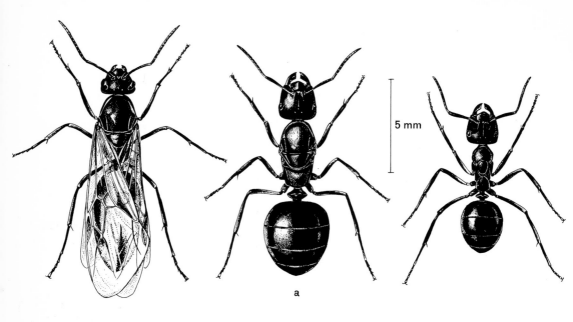

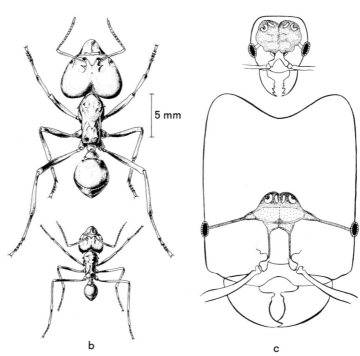

Fig. 42 a. The small red wood ant (Formica polyctena). *Left, male; center, female after shedding of wings; right, worker.*
b. The leaf-cutter ant (Atta laevigata). *Top, soldier; below, worker. Same magnification.*
c. The ant Pheidole instabilis. *Top, head and brain of worker; bottom, soldier.*

a

b

c

5 mm

5 mm

ants need powerful masticatory muscles and, hence, a large head capsule to accommodate them. No increase of mental capacity may be inferred from the size of their heads. The brain of soldiers is in no way enlarged (fig. 42 c).

Caste specialization may go very far indeed. Some species of the genus *Colobopsis* build their nests in tree trunks and connect them to the outside world by a tiny hole which is only just large enough for one ant to pass through. Their community includes a not very numerous caste whose entire mission in life is to act as doorkeepers. They have enlarged heads, flattened in front, that fit exactly in the entrance hole so that they can function as live plugs. Moreover, the texture and color of the head, as far as it is visible from the outside, is such that it can hardly be distinguished from the surrounding bark (fig. 43). A doorkeeper will sit for hours in the entrance hole. She admits only members of her community demanding entrance by taps with their antennae, and these only if she can also recognize their smell. Should the hole be slightly larger than the head of the doorkeeper, the ants use a substance like papier-mâché to narrow the entrance until it fits the head exactly. When the opening happens to be exceptionally large, several doorkeepers may block it jointly (fig. 43, bottom).

The employment of gatekeepers to guard buildings is an age-old human custom. But the ants are unique in having gatekeepers that block the entrance with a strongly armored part of their own body, camouflage it at the same time, and let no one enter without the correct password.

In temperate latitudes one can sometimes watch a mighty swarm of winged sexual forms rise from an anthill on a fine summer's day. But this is quite different from the "swarming" of the bees. We have seen that when bees swarm, the whole colony divides: the old queen leaves the hive with about half the worker bees, and those remaining adopt a new queen. But when "flying ants" leave their nest in a similar frenzy, the swarm consists entirely of young males and females that embark on their nuptial flight at a propitious hour and try to achieve a union. The males die soon after the flight. The females shed their wings at a preformed abscission point near the wing base because they will not need them again. They try either to found a new colony or to join an existing one.

When an ant queen founds a new colony, she is entirely on her own. She searches for a hiding place or digs a small

Fig. 43. Top, entrance to an ant colony (Colobopsis) *in a tree trunk, blocked by the head of a doorkeeper. A worker is demanding admission. Below, a larger entrance hole is kept blocked by a group of doorkeepers.*

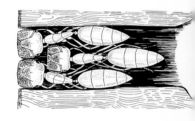

cavity and then blocks the entrance from the inside. There she will spend the next few months—in some species even more than a year—in complete seclusion, as if in solitary confinement. She lays a pile of eggs and tends the brood. During this period she lives almost entirely on the reserves contained in the strongly developed flight muscles of her thorax which are now redundant and gradually degenerate. She feeds her brood with a secretion of her salivary glands and with surplus eggs. Occasionally she allows herself the luxury of eating one of these. When the first workers emerge from their pupae, they open the chamber, fetch food and building material, and take over all the other chores. The new ant community begins to function. Most species of ants found new colonies in a similar way. Their behavior in this respect is reminiscent of that of the bumblebees.

Alternatively, the fertilized queen returns to her own nest where she is made welcome, or enters another nest of her own species. This is how colonies with several, and sometimes many, queens develop. Such colonies may found daughter nests in the vicinity, each with its own queen, but remaining firmly connected with the mother nest through permanent ant trails. Occasionally one finds extensive settlements with several "satellite towns," all belonging to the family of the original colony.

In contrast to worker bees, worker ants may live for many years. Their queens may live to a ripe old age. The Swiss ant specialist Henri Kutter once kept a queen of the common brown garden ant (*Lasius niger*) in an artificial nest and observed her for almost twenty-nine years. She laid eggs until she died. Life spans of fifteen to twenty years have been observed in other species more than once. Since queens do not mate again after their nuptial flight, they have to store in the sperm pouch of their abdomen an incredible number of live sperm cells. Three hundred and twenty millions were found in a queen of the leaf-cutter ants.

The queens do nothing but lay eggs all their lives. The workers are engaged in a multitude of different tasks. Their principal duty is the care of the brood. The larvae of ants, like those of bees, are helpless grubs that have to be fed and tended. Just before they pupate, they can spin a cocoon like the silkworm and spend their pupal stage in this silk casing. Whereas bees spend their early life until the end of their pupal stage in one and the same waxen cradle, the eggs, larvae, and pupae of the ants may be found in neatly sorted piles in all chambers of the nest (fig. 44). There is no trace here of cells precisely adapted to the size of the larvae nor of a meticulously regulated temperature in the brood-raising area, as with the bees. But the ant brood is by no means neglected. Because it has the advantage of mobility, it can be carried by the workers to those parts of the nest which, at any given moment, have the most favorable temperature and humidity for its development. The brood is treated as the colony's greatest treasure. Anyone who, for some reason, has had occasion to disturb an anthill will have been impressed by the lightning speed with which workers pick up the brood and carry it to safety.

The workers are also charged with the defense of the nest. Like bees and wasps, they too are well armed and possess a poisonous sting. Admittedly, this is no longer functional in some species, but the poison glands remain; instead of injecting the poison, these ants squirt it into the wounds which they have inflicted with their pointed mandibles. This same equipment enables them to lead a successful predatory existence. Many ants live on other insects which they attack and drag to their nests, alive or dead. But their diet is varied and differs between species. Some prefer the sweet honeydew exuded by aphids and coccids, and others live on fungi which they cultivate in their nests. But ants of the temperate zones never hoard food for the winter, and their capacity for regulating the temperature in the nests is limited. Nevertheless, they manage to survive the winter. They withdraw into the depths of their nests and remain below in a state of torpor induced by the cold. Their metabolism is reduced to a minimum, enabling them to live without food for a period of several months until the warm rays of the spring sun call them back to active life.

Obviously, these brief remarks by no means exhaust the variety and complexity of ant life. But I fear the reader will now be impatient to hear about their architectural achievements. These are as varied as the mode of life of the builders themselves. This great variability of their behavior patterns may explain why the ants have colonized practically the whole globe. The honeybees surpass the ants in physiological achievement and social organization. Why then, one might ask, do ants, but not bees, occur virtually everywhere where terrestrial life is possible? Why do a multitude of genera and species inhabit all five continents in numbers never approached by

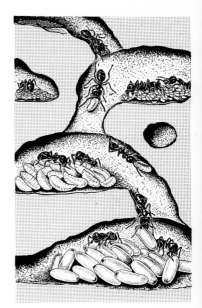

Fig. 44. Detail from a nest of the common brown garden ant (Lasius niger). Brood tended by workers. Top, eggs; center, larvae; bottom, pupae in their cocoons.

other insects or vertebrate species? The reason is that, in the course of their phylogenetic evolution, the ants have been able to develop suitable modes of life for the most diverse environmental conditions. In turning now to their structures, I must restrict myself to a few selected examples.

From modest soil dwellings to stately mounds. The nest of *Myrmecia dispar,* an Australian species belonging to the most primitive group of ants, consists merely of passages and cavities dug into the ground. Nests of this sort are not difficult to build and would appear to rank no higher than the burrows of digger wasps and similar nesting cavities as an achievement. But, in fact, the demands that a social community such as an ant colony makes on its home are considerably greater than those of a solitary insect. The various excavations must follow a definite plan and serve a variety of purposes. In a young nest (fig. 45, left), the lowest chamber is the sheltered place where the brood is housed and lovingly tended by the queen. When the first workers emerge, they extend the structure to greater depths and build horizontal galleries radiating from the main shaft in all directions. Normally, the deepest cavity continues to serve as the main breeding chamber, but in periods of rapid population growth some higher-lying "residential chambers" may also have to be used to

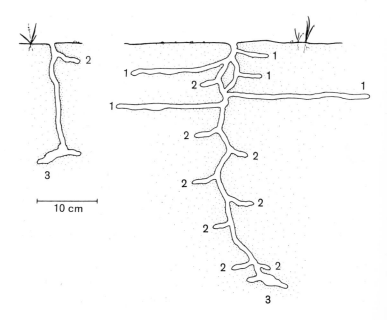

Fig. 45. Primitive soil nest of an Australian wood ant. Left, young nest; right, older nest. (1) Refuse depositories; (2) living chambers, partly used for brood; (3) main brood chamber.

10 cm

some extent for the brood (fig. 45, right). The passages closest to the surface are used for refuse. They are packed tightly with empty cocoons, parts of the dead bodies of ants, and the uneatable remains of prey. Nests of this type are difficult to find because all that is visible from the outside is a small hole through which ants run in and out.

What a contrast to the imposing hills built by the large red wood ants (*Formica rufa*) which are widespread over Europe and Asia. Their inhabitants are not numbered in hundreds like those of the small Australian soil nests, but in hundreds of thousands. They have chambers not only in the visible mound but excavation may reach down as far below ground as the mound rises above it (pls. 53 and 54, p. 98). To strengthen the walls of their subterranean passages and cavities, these ants usually employ a kind of mortar which is a mixture of soil and secretions from their own bodies. Nests are always started below ground. For instance, when an overpopulated nest starts a daughter colony, a group of workers will first dig cavities in the ground at a suitable site and link them with passages. Or the nucleus of the new nest may be a tree stump, the workers hollowing out the first chambers and passages from the decayed wood. But soon there is the urge to cover up and hide all openings leading to the outside world. Pine and spruce needles, small twigs, bits of moss and lichen, grass stalks, and the like are brought along. This is the start of mound construction which from now on will be vigorously pursued simultaneously with the extension of the structure underground. Pieces of light material, such as fir needles, are picked up and carried home at considerable speed, held aloft in the mandibles. Twigs and other heavy pieces are dragged along the ground, and where a load is too heavy for one worker, other nearby ants will join in and help with the task. Not infrequently progress is slow because they all pull in different directions—though they all want to get to the same place and have the capability of using scent trails, landmarks, and the position of the sun for orientation. But one must not forget that keeping to a given direction when maneuvering an awkward load, especially over ground covered with vegetation, is much harder than direction-finding in free flight. After much pulling this way and that, they finally arrive at the nest. Only some of the material is carried to the top of the mound. The greater part is deposited on the slope. In this way, the nest receives its regular rounded shape (pls. 53 and 54, p. 98). Smaller particles are used to ensure a dense surface and to line the inside of the various chambers and connecting passages. The surface has a great number of holes which serve as entrances. These are not permanent. Their number and position is often changed. At night and on cold days the ants plug them with nesting material just as we close our windows to keep the heat in. Though an ant heap may appear to be a finished edifice, building work goes on all the time and is not confined to the opening and closing of entrance holes. The whole mound is in a constant state of re-arrangement and reconstruction. Werner Kloft once sprayed a blue dye, of a kind that dries quickly and does not make particles stick together, over a mound of the small red wood ant (*Formica polyctena*). Four days later the mound was brown again, and the blue particles lay eight to ten centimeters below the surface. In the course of a month they had sunk down to a depth of forty centimeters and were more loosely distributed. Other dyes sprayed on later went exactly the same way, and four weeks after application all colors began to reappear on the surface in the original order. The explanation for this is that the ants constantly carry nesting material from lower strata to the surface and deposit it on the outside of the nest, causing the original surface layer to move gradually down until eventually it becomes the turn of the components of that layer to be carried up again. The biological significance of this process is clear. This constant turnover causes the material from the humid interior to be regularly dried on the surface, and this helps to prevent mold formation. Abandoned ant heaps are soon subject to fungal decay and breakdown.

This process is at the same time important for the stability of the mound and for its protective function. Large unwieldy pieces, such as small twigs or long pine needles that cannot pass through the rest of the building material, will not be brought to the surface again. They constitute the solid core of the heap, while the fine particles, when packed tight, form an excellent outer layer and give protection against wet and cold. Another duty of the workers is to keep the slope of the mound at the right angle. In this they are to a certain extent guided by the local climate. Mound-building is an important method for obtaining sufficient solar heat in temperate latitudes. In the late and early hours of the day when the sun is low on the horizon, a domed nest intercepts more solar rays than could otherwise be utilized for the heating of the nest (fig. 46). For social insects lacking the highly developed systems of temperature regulation of the bees and wasps, this is an effective, if simple, way of heating. Ants often gain further warmth for their nest by acting as live heaters. They sit in the sun in large numbers, and when they are warmed through, they take their hot bodies down into the nest while others come outside to repeat the process. In contrast to the bees that create the right temperature and humidity conditions for their brood in a special breeding area, the ants achieve largely the same effect by carrying their brood to those parts of the nest where conditions are most favorable at a given moment.

Wood ants are not the only mound-builders among the ants, but the mounds of the others are smaller. The common brown garden ant and the small yellow meadow ant (*Lasius*) build their modest little mounds of earth using grass and other plants as support. Their nests, too, extend deep into the ground below (fig. 47). *Formica* and *Lasius* species build anthills of a similar kind in North

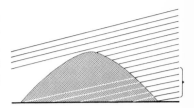

Fig. 46. How an anthill catches the rays of the sun. If there were no mound, the nest would only be warmed by the rays shown in white while the sun was low on the horizon. The mound makes more heat available.

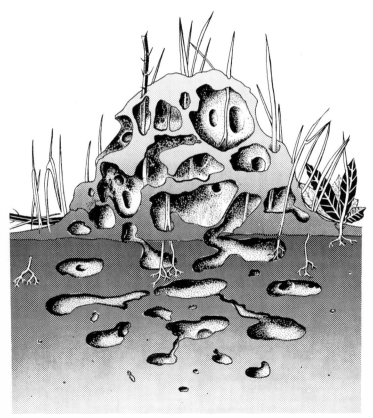

*Fig. 47. Cross section through nest of the brown garden ant (*Lasius niger*). The chambers reach far into soil below nest.*

America and Asia, and there are other mound-building species in other continents.

Dwellings made of wood or paper. Tree trunks are favored nesting places. The *Colobopsis* species, which have a preference for homes carved out of walnut trees, have already been mentioned. There they can live in undisturbed safety, secluded from the world outside, their entrance hole blocked by a doorkeeper. A carpenter ant, *Camponotus herculeanus,* prefers stems that are partly decayed and therefore easier to tunnel in, but sometimes it builds in sound wood. In this case, the layout of the nest is determined by the structure of the tree's annual rings. Each of the concentric layers in the wood of a tree consists of a lighter colored, comparatively soft inner section laid down during rapid growth in spring, and a darker, more compact outer section of "summerwood." When constructing their chambers, these ants attack the softer parts of the annual rings and carry away the wood particles. The hard parts are left in place. This gives them the double advantage of easier work and stronger walls (pls. 55 a and b, p. 99).

Not all species hollow out their nests by gnawing the wood away. Some build them freely into an existing cavity. The nest shown in plate 56 (p. 100) was built by the jet ant (*Lasius fuliginosus*) in a hollow tree trunk. It possesses a subterranean part in which the ants hibernate. The walls of the many irregular chambers in the tree cavity itself consist of a kind of paper pulp ("carton nests") which the ants produce from wood particles (pl. 57, p. 100). They, too, must be given credit for the invention of paper. But the process they employ is quite different from that of the wasps, at any rate, from that used by the *Lasius* species studied by Professor Bert Hölldobler. Whereas the wasps bind their wood fibers with saliva, the ants impregnate their particles with a highly concentrated sugar solution to make them stick together. This sugar solution serves at the same time as a nutrient substrate for a fungus (*Cladosporium myrmecophilum*) that is always found in the nest of these ants but nowhere else. The interwoven strands of the fungal hyphae give strength to the light structure, and there is thus a symbiosis between fungus and insect giving advantages to both parties. The fungus has a sheltered place to grow in; the ants have a strengthened dwelling. But where does the sugar come from? I shall have more to say later about the habit of many ant species of collecting the sweet exudations of aphids and coccids for food. *Lasius fuliginosus* belongs to this group, but with this species, as we have seen, the sugary liquid is not only an important source of food for the ants themselves but serves at the same time as a nutrient substrate for the fungus which they cultivate as a kind of live mortar for their structures.

Close observation of the building process revealed an interesting division of labor. One group of ants brought the wood particles and deposited them on the building site. Others brought the sugar solution in their crops (a dilatation of their anterior intestine) and distributed it. A third group, the actual builders, picked up the wood particles with their mandibles, moistened them with sugar solution from their crops, and stuck them on the growing wall of the chamber.

Lasius fuliginosus is not the only *Lasius* species that builds carton nests. Such nests are built not only in cavities but may be found, for instance, in the angle between roof timbers in an attic. Though few European ant species build them, carton nests are widespread in the tropics, especially on the American continent, India, and Madagascar. Since cold is not a problem for the ants in these regions, it is not surprising that such nests occur in places without shelter, maybe high up in the crown of a tree or attached to a tall stem. Their size and appearance may be similar to that of large wasps' nests. Nothing seems to be known about the working methods of tropical papermaking ants.

Ants as weavers. In the crowns of trees in tropical southern Asia, one occasionally comes across the round or oval leaf nests of a reddish, fairly large species of the genus *Oecophylla*. They consist of living, undetached leaves held together by a dense silky web (fig. 48). For zoologists, these nests were once a major puzzle because ants possess no spinning glands. Admittedly, their larvae have spinning glands and, in many species, spin cocoons of silk strands before they pupate like silkworms and many other caterpillars. But this fact alone could not solve the problem because ant larvae are helpless grubs tended by the workers in the depth of the nest, and it would be impossible for them to crawl to the surface and join the leaves together with threads. Working on the assumption that the ants would reveal their secret if they could be induced to come out to repair a damage deliberately in-

Fig. 48. Nest of the weaver ant
(Oecophylla smaragdina).

flicted on their dwelling, a brave naturalist once climbed
up into the crown of a tree and made a slit into such a
leaf nest. He had to be brave because he knew that these
ants vehemently attack intruders and seek to drive them
away with painful bites and squirts of corrosive poison.
In this case they tried in vain, and his bravery was rewarded, for soon a group of ants came along and took up
a position on one side of the tear. Attaching themselves
firmly with the sharp end claws of their six legs, they
seized the other edge with their mandibles and carefully

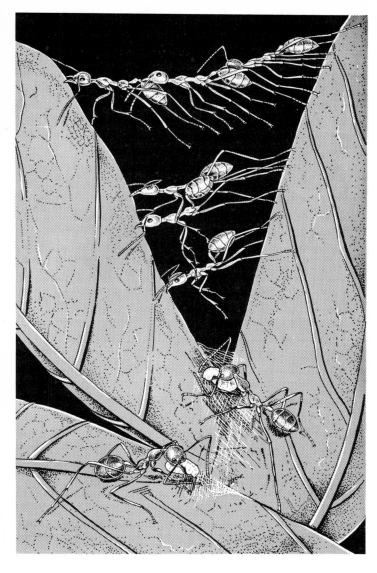

Fig. 49. Weaver ants at work. Center, workers trying to pull leaf edges together. Top, where edges are too far apart, ants form living chains. Bottom, workers weaving leaves together; each worker holds a larva between her mandibles, using it both as spindle and shuttle.

tried to pull it closer by moving their legs, one after the other, further back. It was a strange sight. While the gap gradually got smaller by their joint efforts, other workers appeared and carefully cut away the ragged ends of the torn web with their mandibles. They carried them to an exposed part of the nest and, opening their jaws wide, let them be carried away by the wind. Where the distance between the two edges was too great, other ants were seized to form a living bridge across the gap. What happened next was even more astounding. A group of workers, each carrying a full-grown larva, emerged from the depth of the nest. Where the two edges had been pulled together sufficiently, they went to work pressing the mouths of the larvae (which were thus compelled to act as live shuttles) against the leaf surface first on one side, then on the other, and, by squeezing them with their mandibles, made them discharge some of their glandular secretions (fig. 49). In this manner the two edges of the tear in the nest cover were stitched together again with the silken threads produced by the larvae. This use by the ants of their own larvae as both spindle and shuttle is probably the most remarkable example among the few instances of the use of tools by animals.

We owe this interesting description to Franz Doflein, a Munich biologist. When he arrived in Ceylon on a study tour through the tropics, the woven ants' nests fascinated him to such a degree that he was determined to get to the bottom of the mystery. He succeeded on the very last day of his stay in Ceylon. After his return home, he made a discovery which, alas, many discoverers have had to make. On studying the literature he found that an Englishman, Ridley, had seen and described the nest-building process before him. Still, it does not really matter if two people independently experience the same rare thrill of discovery.

For our illustrations of the nest (fig. 48) and its inhabitants at work (fig. 49) we are indebted to Mrs. Turid Hölldobler, who was able to observe the weaver ants when she and her husband visited Ceylon in 1972. In Doflein's illustrations, the ants, carrying the larvae, are shown in a straight line along the edge like soldiers on parade, and the silk threads laid across the gap are parallel. Hölldobler watched the web taking shape by less regular movements, but the principle was the same. Quite possibly small differences in behavior do occur.

Differences between species are much greater. It is now known that other tropical ants (*Polyrhachis* and some species of *Camponotus*) similarly use their larvae to produce silk webs. But their nests do not necessarily resemble each other.

Some only join a few leaves together with loose threads to give them shelter; others fashion dense webs with several chambers. Some nests consist of a tightly wrought bag of silk attached to a leaf. The nest of yet another species was freely attached to a tree trunk covered with lichen. It was almost impossible to see because small particles of lichen, bark, and other matter had been woven into the silk mat, camouflaging it to perfection. The fact that the art of nest-weaving is widespread and occurs in many variations makes it even more enthralling. We should dearly like to know how the ants came to develop this strange building technique. But we have not even begun to penetrate their secret.

And how about the fate of the larvae that have to give up some of their silk in this way? Are they still capable of spinning cocoons for themselves? In some cases the answer is easy. The larvae of *Oecophylla* and *Polyrhachis* never spin cocoons and their pupae lie in the nest without any cover. Their silk is used in nest-building only. But this is not universally so. Friedrich Schremmer told me personally that he found in Colombia a weaver ant of the genus *Camponotus* (presumably *Camponotus senex*) whose larvae spin cocoons. Whether they can still do so after they have been pressed into service to supply silk for building is something which, so far as I know, has not yet been investigated.

Storage chambers and culture rooms. The examples I have selected show how varied are the materials used by ants in the construction of their dwellings, but interior layout is more uniform. Nests consist chiefly of a number of chambers connected by tunnels and passages. The chambers serve mainly as living quarters for the ants and their brood. In some species storage chambers play an important role. I have mentioned before that ants do not hoard food for the winter—indeed this would be of little use as the temperature regime of the ants precludes any activity in the cold season. But in regions of the tropics and subtropics where drought rather than cold periodically threatens survival, some ants do accumulate stores. Seed-storers are well known in this respect. During the period of plant growth and seed ripening, they collect great quantities of seeds and store them in their chambers against the approaching long period of drought. The European harvester ants of the genus *Messor* in the Mediterranean region and the harvester ants (*Pogonomyrmex*) of North America belong to this group. Their nests are dug several yards deep into the ground. In some North African species a single settlement may extend over an area of fifty meters in diameter. A sizeable part of the country's grain crop disappears into this subterranean world.

The storage chambers of the desert ants that live in the arid regions of North America are completely different. The ants of the genus *Myrmecocystus* collect sweet plant juices, especially the sugary exudations of oak-apples, which are a highly nutritious form of food. But how do ants that lack the skill to construct completely tight containers go about storing the sugary liquid? The method they contrived is most amazing: they feed a number of workers with the sugar juice until their abdomens swell like balloons and the chitinous plates of their abdominal segments are widely separated by bands of extended connective skin. These living storage vats hang from the ceiling of the storage chambers, looking very much like little Chinese lanterns hung up by gnomes for some underground festivity (fig. 50). Only they do not glow in the dark night of their prison but wait, motionless, until, in the months of shortage, the food-gatherers among the ants collect the sweet contents of their stomachs drop by drop from their mouths.

Instead of collecting stores, the leaf-cutter ants of South America (for instance, those of the genus *Atta*) cultivate their own food in special chambers of their wide-ranging subterranean colonies. To prepare the compost beds for the cultivation of their fungus, they have to cut pieces of leaves from the trees in the vicinity. Because of a preference for the tender leaves of orange trees and other cultivated plants, they can cause devastating damage. In endless columns they climb up the tree trunks and

Fig. 50. The American desert ant Myrmecocystus *employs some of its workers as living honey jars. These are filled almost to bursting point: the widely separated chitin plates of their abdominal segments are connected by stretched skin. Worker at the top right (she is 7 mm. long) collects food at one of the "honey jars" (length 15 mm.).*

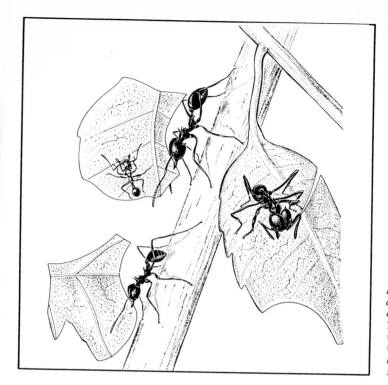

Fig. 51. Leaf-cutter ants. The ant at the right is just cutting off a piece of leaf with its sharp mandibles. The others are on their way home with their loads. Top left, a small worker rides home on a piece of leaf and repels the attacks of parasitic flies with wide-open jaws.

spread all over the branches. The workers of the genus *Atta* vary greatly in size. The largest are the soldiers, the defenders of the colony. Workers of medium size cut off pieces of foliage (fig. 51, right). Holding their abdomens well tucked in, they busily make curved incisions with their sharp mandibles, seize the severed parts, and set off with them for home (fig. 51, left). When the American ethologist Donald R. Griffin studied these ants in Trinidad, he noticed that on their expeditions the leaf cutters were accompanied by some extremely small workers that watched the operations carefully. Just before a piece of leaf was completely severed, they climbed on to it and allowed themselves to be carried home in style (fig. 51, top). Dr. Irenäus Eibl-Eibesfeldt succeeded in unraveling this curious behavior. The leaf-cutter ants are exposed to the attacks of certain small parasitic flies which lay their eggs on the necks of the ants. When the fly larvae hatch, they burrow into the ants' heads and devour the contents. The small ants successfully repel the approaching flies by awaiting them with wide-open jaws. There is thus a biological reason for the fact that not only soldiers but also extremely small individuals can show a strong,

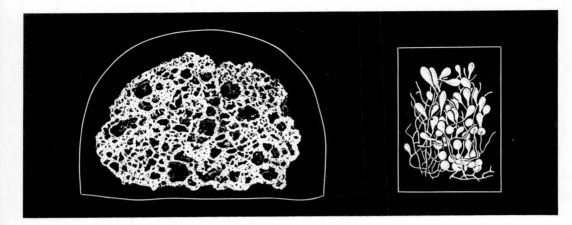

Fig. 52. Left, fungus mass from the nest of a leaf-cutter ant (Atta texana). Right, a small part of the fungal mass, greatly magnified. The clublike swellings at the ends of the mutilated hyphae form the exclusive food for the whole colony.

if highly specialized, defensive instinct. A large ant sitting on a leaf would be too heavy to carry. Not only do they function as defenders against air attack, however; the main task of the small workers of the colony consists in tending the fungus gardens.

In spacious chambers, which may be up to one meter long and thirty centimeters wide and high, the leaf fragments are chewed into small pieces, mixed with saliva, and fertilized with a few feces. This is the material of the spongy lumps in which the fungus is cultivated by the ants (fig. 52). Thanks to the care lavished on it by the workers, it grows extremely well, partly, no doubt, because it is treated with a glandular secretion that suppresses the growth of bacteria and other fungi, as described by Professor H. Schildknecht in 1971. The ants snip off the ends of the vigorously growing hyphae which, because of this mutilation, develop club-shaped heads (fig. 52, right). These form the only kind of food consumed by the colony. Both parties profit greatly by this symbiosis. The fungus grows in a protected environment, lovingly tended by the ants. In exchange, it breaks down and utilizes the indigestible cellulose of the leaves carried home by them, thereby turning it into easily digestible food. Well-tended fungus gardens never develop fruiting bodies. Such bodies do, however, very occasionally occur in abandoned parts of a nest, and they have enabled us to recognize the fungus as a basidiomycete. It belongs to the same group of fungi as many of our edible mushrooms, but it is a separate species which is never found outside the nests of ant colonies.

Such nests may reach down to a depth of five meters with their chambers and, in some cases, may have more than a million inhabitants. They are the descendants of one single queen who, when she founded her nest, took with her in a pocket of her mouth cavity a small piece of the fungus mat of the mother nest as her most precious possession. This fungus strain is passed on from generation to generation.

It is a remarkable fact that ants may be growing fungi for entirely different purposes: either as a food or as a means of strengthening their nest walls (see p. 110 f.).

Building roads and animal sheds. The fungus-growing leaf cutters are the only ants that are strict vegetarians. Many species prefer a mixed diet. They hunt insects but also collect sweet plant juices—I mentioned this habit when I described the honeypot ants of the American desert. They do not usually collect their sugar directly from plants as the bees do when they collect nectar from flowers. As a rule, they obtain it from coccids and aphids. Large aggregations of aphids (plant lice) can often be seen on young shoots in which they have embedded their stylets. Similar large aggregations on branches and tree trunks are often formed by coccids (scale insects), a closely related group; these, however, are rather inconspicuous because their bodies are hidden under a scale that adheres closely to the bark. The sap these insects obtain from the plants is not rich in proteins but rich in carbohydrates, of which they exude the unused surplus in the form of sugar. The ants make use of this habit and are careful not to do any harm to the providers of this sweet bounty which otherwise they might consider delectable morsels for immediate consumption. Instead, they stroke them with their antennae, and the little insects respond by offering the ants a sweet droplet from their anus (fig. 53). Frequently, however, they splash sticky fluid in-

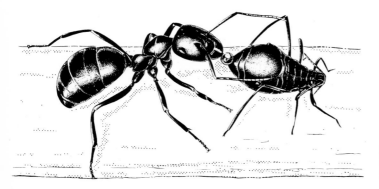

Fig. 53. A sucking aphid is stroked by an ant with her antennae and releases a drop of honeydew.

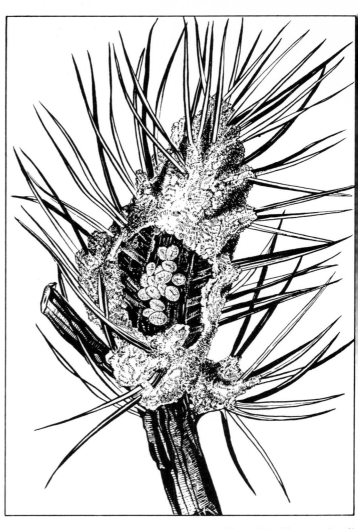

Fig. 54. Ants (Crematogaster pilosa) *have built a carton tent over a colony of coccids. A hole has been made in the tent wall to show the coccids sucking on a pine shoot.*

discriminately about, and the ants collect this "honeydew" from the foliage (the bees often do the same).

In itself this has nothing to do with building. But sometimes the ants build a kind of cattle shed for their "milch cows" by surrounding parts of a shoot inhabited by a colony of aphids or coccids with a roofed enclosure made of soil particles or other material. Garden ants (*Lasius* species) even build little pavilions over the colonies of plant lice or scale insects visited by them. A carton tent built by the North American ant *Crematogaster pilosa* over a colony of scale insects on a pine shoot is shown in figure 54. Doflein observed in Ceylon that close to the main nests of weaver ants there were small nests that contained neither brood nor sexual castes. These were found in places of large aggregations of scale insects, their only inhabitants. These small nests belonged to the main nest as dairy buildings belong to a farm. Worker ants visited them regularly to collect the sugary liquid. Some ants keep and tend scale insects sucking on the roots of plants inside their subterranean dwellings, and sometimes their keepers build special roofed shelters for them.

The closest, and most remarkable, association between ants and coccids occurs in some species of the genus *Acropyga*. *Acropyga maribensis* is the most common ant species in the coffee plantations of Surinam (Dutch Guiana) in the tropical northern region of South America. The ants build their subterranean nests in the root zone of coffee plants. In one respect their activities are good for the plantations because they loosen the soil, but in another, they present a danger because their nests contain a scale insect which sucks at the roots of coffee trees and can cause serious damage. This particular coccid species is only found in the nest of these ants, and no *Acropyga* nest is without them. The symbiosis is a very close one. The ants tend their sugar suppliers like their own brood. The most remarkable behavior in this context is that of the swarming female. While the males at swarming time simply fly away, the winged female before doing so carefully picks up a little scale insect, which must be in a particular stage of development, namely, young but already mated. The queen takes it with her on her nuptial flight, holds it firmly between her jaws while she unites with a male in the air, and brings it back down to earth where she starts looking for a crevice near a coffee plant in which to build her nest. Eventually she deposits her living dowry on a coffee root into which the scale insect at once inserts its stylet. The production of sugar is under way. Once she has her mandibles free again, the queen digs a small cavity into the soil and closes it against the outside, to be the nucleus of her new nest where she will lay and tend her eggs. Should she be disturbed during this work, it only takes her a few seconds to pull the stylet of the scale insect out of the root it sits on before she carries it to a more sheltered place. The behavior of the *Acropyga* queen explains why coccids are found in all colonies of that species. There is an interesting parallel between *Acropyga* who makes sure of the future food supply of her colony by taking with her a coccid capable of propagation, and the leaf-cutter queen who takes away a piece of the fungus garden from the old nest to start up new beds in her own colony. The symbiosis between certain species of ants and plant lice goes back a very long time. Ants along with their plant lice have been found in amber from the early Tertiary period. Neither have changed much in appearance since their ancestors were suffocated in the viscid resin exudations of trees some forty million years ago, and their association, too, has been a lasting one. But we have no way of telling whether the ants in those bygone times built sheds for their "domestic cattle."

Large ant colonies do not limit their foraging expeditions, be they hunting forays or visits to plant lice colonies, to the immediate vicinity of the nest. Their destination may be forty, fifty, even sixty meters away. The habitat of meadow ants (*Formica pratensis*), with its densely growing grasses and herbs, is as full of obstacles for animals of their size as a jungle would be for us. These obstructions are particularly frustrating when large quarries have to be transported to the nest. This explains why these species and many others engage in *road building*. It is not unusual to find half-a-dozen trails radiating in all directions from a meadow ants' nest. The insects remove all grasses and other stalks over a width of about four centimeters, and dig a trench one to two centimeters deep. Thereafter, traffic can move smoothly, though ant roads, even more than our own, require constant maintenance for grass grows again very quickly. Because major obstacles whose removal would overtax the strength of the ants have to be bypassed, trails often meander through the meadow but never lose sight of their goal. Smaller trails, used by some of the ants to gain access to nearby hunting grounds, branch off from the main trunk roads. Sometimes a closed column will march along such a road over long distances, for instance, when leaf-cutter ants visit a particular tree to harvest its leaves, escorted by formidable and very aggressive soldiers. Communications and the safety of travelers were a problem even before the advent of man. The ants appear to have found a reasonably satisfactory solution.

Vagrants without fixed abodes. Having talked so much about the homes of ants, I am almost afraid to admit that some species have no fixed abode but roam the country like gypsies. I take as my example the South American genus *Eciton*. The ants of this genus form enormous colonies but never build nests. They only establish bivouacs. Prolonged observation reveals a regular alternation between two types of behavior. There is first a restless migratory phase in which the ants move their camp every night by a few hundred yards. The queen accompanies her people on the march, and the brood is carried along. It is followed by a stationary phase in which the same camping site is used for several weeks. This curious alternation in behavior is caused by the queen's habit of laying her eggs in batches, separated by periods of inactivity. In communities of this size, the amount of food needed by the larvae once they have reached a certain stage of development becomes enormous, and this need starts off a migratory phase. Each night the whole colony moves to a new camp in some soil cavity or other sheltered place. From there, a mighty stream of millions of ants (few only staying behind in the camp) pours over the countryside in search of new, unravaged hunting grounds. Their pointed jaws kill every living creature they can overpower. The next night they march on. Once the larvae have pupated food requirements suddenly drop sharply, and smaller forays suffice to provide more than enough. The queen benefits from the abundance and a new spate of egg-laying results. The bivouac can now stay in the same place for some weeks. But even for these rest periods the ants will not build a nest. One might say that the ants themselves, the whole colony, *is* the nest, as they hang from the ceiling of some cavity in an enormous cluster, surrounding and protecting the queen and her brood in the center.

Termites, masters in building and civil engineering

There are more than two thousand species of termites living in tropical and subtropical regions. They do not occur in most temperate latitudes (other than North America), which is a good thing, for the damage they do to wooden structures can be devastating. It is not unusual for buildings to collapse without any prior signs of the infestation that has destroyed their timbers. The only people who regret that the homes of these interesting creatures are so far away are European biologists. For centuries scientists have been prepared to undertake arduous journeys to study termites in their homelands.

Termites are often called "white ants." This name is doubly misleading. Despite certain obvious similarities with ants, such as the large numbers of wingless forms which occupy the nests and the appearance from time to time of swarms of winged sexual forms, both groups are entirely different morphologically and phylogenetically. And while it is true that most termites are white because they usually live in the dark, some species do not shun the light and are as dark as ants. Unlike the bees, wasps, and ants which all belong to one insect order, the Hymenoptera, and therefore are closely related, the termites belong to the order of Isoptera and their closest relations are the roaches, which include the cockroaches of our kitchens: this group of insects is much older and more primitive. There is no doubt that the termites developed their social organization independently from the honeybees and the other social Hymenoptera. There are astounding parallels, but also completely different solutions of similar problems. As far as building is concerned, their achievements stand unique.

The pathways by which termites and termite communities developed from their cockroach-type ancestors are not known. These events took place millions of years ago, and no transitional forms between solitary and social species, such as survived among bees, are in existence today. All known termite species, like all ant species, are social insects. Their colonies may have over ten million individuals, thereby surpassing even the largest ant communities. As before, I intend to describe briefly the appearance, social organization, and mode of life of these insects before I pass on to a description of their buildings.

The termite community.   Most termite species avoid the light. Their nests are in the soil, in wood, or inside tall mound-like termitaries, and they tend to use subterranean passages or covered galleries for their runways. Termites have a tender skin and cannot withstand desiccation. They can thrive only where there is warmth and high humidity. The fact that they dwell in darkness explains why they are blind or have only rudimentary eyes. The eyes are well developed only in the sexual forms. Males and females look very much alike. Both have well-developed wings (fig. 55, p. 126), though they keep them only a short time. Once or twice a year, young sexual forms appear and, like those of ants, swarm in enormous numbers, leaving their dark nests for the brightness of day. But whereas sexual ants copulate during the nuptial flight and the males have no further function, the swarming of termites leads only to a kind of "betrothal." Soon after they return to earth, they shed their wings, which will henceforth no longer be needed, and form pairs. The male runs after its female in what is called a "courtship promenade." It may last ten to twenty minutes, or up to two days. During that period, these succulent insects are easily caught and avidly eaten by many insectivorous species and, in some parts of the world, by man. (Roasted termites are supposed to taste better than shrimps.) Only few achieve their goal, which is the construction of a hidden chamber as a nucleus for a new colony. And only then do they reach sexual maturity and enter, as king and queen, into lifelong matrimony. At the beginning, it is they that tend the brood, but soon the roles are reversed. In the later stages, the royal pair is waited on, and devotes itself entirely to the task of reproduction. In highly organized termite communities, the royal chamber and the areas for the young brood around it form the core of the whole nest. In such highly developed species, the king and queen are walled up in a narrow cell where only workers, which are of a much smaller size, can enter and leave through openings just large enough for them. From the point of view of conjugal fidelity, this close confinement would not be necessary since, in a population of over a million termites, neither the king nor his consort could find another partner. In some species, the ovaries develop so enormously that the queen's abdomen is monstrously enlarged (fig. 57, p. 132). In *Macrotermes bellicosus* (the species used to be called *Macrotermes natalensis* but had to be renamed in accordance with the rules of nomenclature), the abdomen may become fourteen centimeters long and three and a half centimeters wide. The number of eggs laid is prodigious—thirty thousand or more in a day. Such a restricted and one-sided life may seem unhealthy, but queens may live for many years—how many is not known. The colonies and their structures are apparently able to outlast centuries, for the sexually reproductive animals can be replaced when they die.

The larvae of ants and other Hymenoptera hatch from their eggs as legless maggots incapable of doing any work. Their larval period is spent in inactivity; they do not participate fully in the life of the community until after pupation and metamorphosis. The development of termites is different. Like cockroaches, grasshoppers, and several other orders of insects, they never enter a pupal stage. Metamorphosis for them is a gradual process: small changes take place at each molt, through which

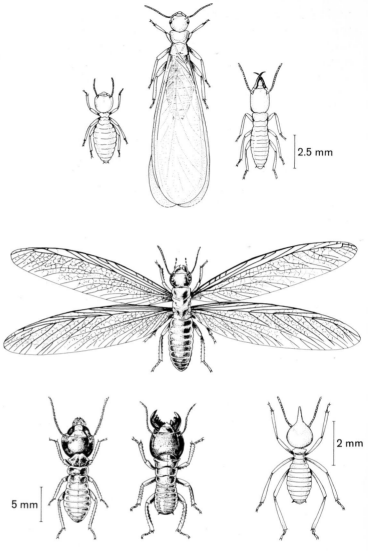

Fig. 55. Some termites of different castes. Top center, winged sexual form; left, worker; right, soldier, all of Coptotermes acinaciformis. *Center, winged sexual form; bottom left, worker; bottom center, soldier of* Hodotermes mossambicus. *Bottom right, nasute soldier of* Nasutitermes exitiosus.

2.5 mm

2 mm

5 mm

the larvae increasingly resemble the fully grown insect. This explains why termites work while they are still larvae. Termite communities, like those of ants, contain several castes differing in morphology and function. Males and females have already been mentioned. The most numerous caste is that of the workers (fig. 55, top left, bottom left). Soldiers occur in all termite communities and, like those of the ants, are usually characterized by large heads and powerful jaws (fig. 55, top right, bottom center), but they may also have other quite fantastic features. In the "nasute" soldiers of certain termite species, the forehead above the jaws is extended into a nose-shaped point (fig. 55, bottom right). Here are the openings of a large gland that produces a sticky secretion. The soldiers cover an enemy with this glutinous substance, rendering him helpless.

Workers and soldiers together are capable of carrying out all the tasks required in a termite colony, though they remain arrested in an adolescent stage in which their sexual organs never develop. The task of increasing the family is efficiently attended to only by the royal pair.

In the communities of honeybees, bumblebees, wasps, and ants, the workers are all females. Their menfolk never work. Not so in a termite state. Both sexually inactive males and sexually inactive females share the duties of workers and soldiers with equal zeal and devotion. There is also a small caste of "supplementary reproductives," stand-ins from which a new male or female develops if a colony loses its king or queen. Hormones play a vital part in such replacements, and in the differentiation of castes as a whole.

Termites, as I mentioned earlier, may become a menace because they destroy wood. Obviously, they do not do this out of wanton destructiveness but because wood is, for many of them, a valuable food. This may sound strange because to most animals, including man, wood is indigestible. But the wood-eating termites live in close symbiosis with certain unicellular protozoan organisms. The flagellates that live and move about with their flagellae in the termites' intestines possess special enzymes capable of breaking down the cellulose of the ingested wood and making it utilizable for themselves and their hosts. This process takes place in a special section of the termites' intestinal tract which has been widened to form a fermentation chamber. The flagellates multiply profusely and supply their hosts not only with digestible carbohydrates but also with the necessary protein, because the surplus population of these small organisms is itself digested in the termites' gut. In one large family, the Termitidae, which comprises three quarters of all termite species, bacteria and not flagellates assist in the digestive process.

The diet of termites is not restricted to wood alone. In some species, it may be quite varied and include all kinds of animal and vegetable matter. It is a remarkable fact that, like the leaf-cutter ants of tropical America, many

Asian and African termites cultivate and tend fungi in special chambers of their nests. Each group has independently developed the same gardening skills.

Even when food is abundant, soldiers are usually unable to help themselves to it. Their heads have become so specialized for other tasks that they have to be fed by the workers.

Simple structures. The nests of the more primitive species are well hidden and difficult to find. They usually consist of an apparently irregular system of passages and chambers which serve both as living and storage quarters. The queen may remain relatively small and mobile and is not confined to a cell like the queen of highly organized termite societies. These primitive groups comprise the species that inhabit dry wood and are the most dangerous pests. When *Kalotermes* species hollow out their galleries in the beams of a building, they follow the structure of the wood in the same way as we have seen carpenter ants do. Some galleries end "blindly" in widened chambers used for depositing feces. This species lives in complete seclusion and makes exit holes only at the time of swarming to enable the sexual forms to leave the nest. The structures of *Cryptotermes,* on the other hand, always possess openings which are used by the inhabitants to get rid of their excreta instead of hoarding it, but these holes are always plugged by the large heads of soldiers—another strange parallel to the behavior of certain ants.

Some of these primitive species build soil nests in prairies and semideserts. *Hodotermes mossambicus* builds spherical nests with many chambers and passages radiating in all directions at depths of three meters or more. The workers of this species collect grass and spread it in special flat chambers situated close to the surface and measuring up to one meter or more in diameter (fig. 56). As soon as fermentation is over and noxious gases can no longer endanger the brood, the hay is carried into storage chambers lying close to the nest.

Most of the termites of North America belong to these primitive forms, as do the few species that occur in Europe: *Kalotermes flavicollis* in the south of France, *Reticulitermes lucifugus* in southern Europe, and *Reticulitermes flavipes.* This last named species was accidentally introduced into Hamburg where for several decades it has survived and multiplied in favorable surroundings, withstanding all attempts to exterminate it.

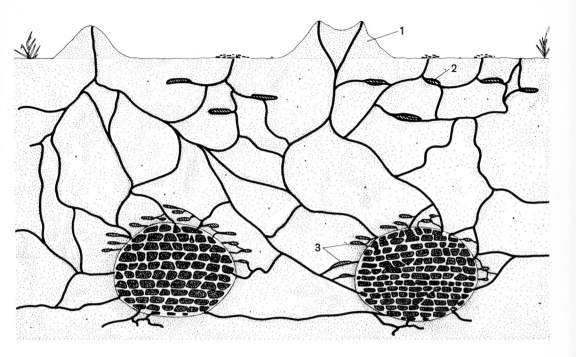

The great architects. The species mentioned so far, and many others, arouse attention by the damage they do rather than by their skill in building. However, termite nests may be gigantic structures, and there are regions where they are so numerous that they determine the character of the landscape (pls. 58 a and b, 60, and 61; pp. 130–31 and 134–35). Some are seven meters high. A number of subterranean passages lead into the surrounding area from which the workers collect seeds, leaves, and other food at night. Life in these enormous hillocks exposed to the tropical sun is possible only because the termites cover them with a compact layer of building material which functions like an outer shell of reinforced concrete and helps to regulate the interior climate of the nest in accordance with their requirements. Where this cannot be achieved completely in all parts of the structure, the termites take great care to lodge their eggs and those larvae which are still young and helpless in places where the interior climate is suitable. Clearly, the strong hard layer surrounding the termitary has other functions as well, in particular the protection of the colony against attacks by the many animals that otherwise might feast on these multitudes of succulent, tenderskinned insects. But even such a formidable barrier does

Fig. 56. Cross sections through two ground nests of termites (Hodotermes mossambicus). Galleries lead from the nests in all directions. (1) A heap of excavated soil; (2) chambers near ground surface for storing freshly collected grass; (3) deep-lying chambers close to the nests for the dried hay.

not give absolute protection against all attackers. Aardvarks smash it with their hoof-shaped claws. The front legs and claws of anteaters are strong enough to break it open. Both animals pull out the termites from their broken nests with long sticky tongues. In plate 64 (p. 136) the small anteater of South America, *Tamandua tetradactyla,* is shown enjoying a meal at a nest of tree termites which hangs from the underside of a branch.

The chambers of termite mounds may extend deep into the ground. All nests, including those of mound-builders, start below the ground. Such nests may stay underground for several years and reach a large size before they become visible on the surface. When the great entomologist Karl Escherich visited Ceylon he had occasion to observe how, after heavy rains, small soil mounds suddenly appeared in areas where there had been no termite mounds before. They had been thrown up in small groups by nests that until then had remained invisible. Their further growth also took place in spurts at intervals of months or even years, whenever the soil was softened and made easier to work by heavy tropical rains. In time the groups of small mounds belonging to the same nest became one sizeable hillock. When they had first appeared,

Plate 58 a (facing left). The structures of compass termites in the Australian steppe. Their broad sides face east and west.

Plate 58 b (above). Their short sides face exactly north and south. The weak morning sun cannot greatly heat the structure. (See pp. 129, 138.)

131

they had looked like anthills. But, in contrast to anthills which are soft and can easily be destroyed by a kick or by a heavy downpour while they are still small, termite mounds are hard and resistant from the beginning.

Now let us break through the forbidding armor of a large termitary and look into the interior. A biologist who penetrates toward the center by carefully removing layer after layer, as if he were slicing a loaf of bread, finds far more evidence of an architectural plan than in an ant's nest. One of the most unexpected features of our present example (pl. 59, p. 133) is the different coloration of the interior. The upper part is dark, the lower, light. This surprising appearance is due to the fact that two species of the genus *Macrotermes* have settled in the same mound, one above the other. Termites of the species *Macrotermes carbonarius,* the occupiers of the upper part, build the walls of their chambers from their own dark excreta. Their bodies, too, are dark, and they are by no means afraid of the light. They can be observed in daylight working in the vicinity of the nest. The light-colored structure underneath is the work of another species of *Macrotermes,* which is pale and shuns the light in the manner of most termites and which uses a pale-colored building material. Both species are fungus-growers, but

Fig. 57. The interior of a royal cell. The abdomen of the queen (1) is enormously enlarged by the number of eggs. Eggs leaving the tip of the abdomen are taken over by the workers. The king (2) sits next to his queen who is tended by large numbers of workers and guarded by soldiers (3).

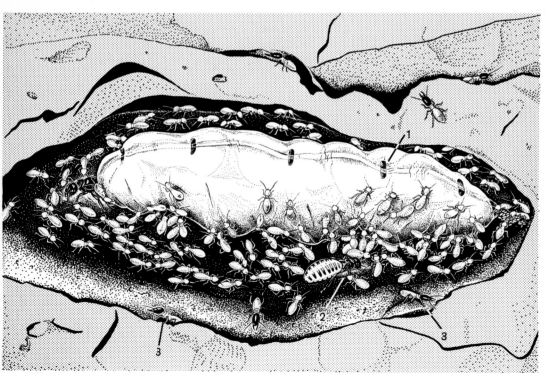

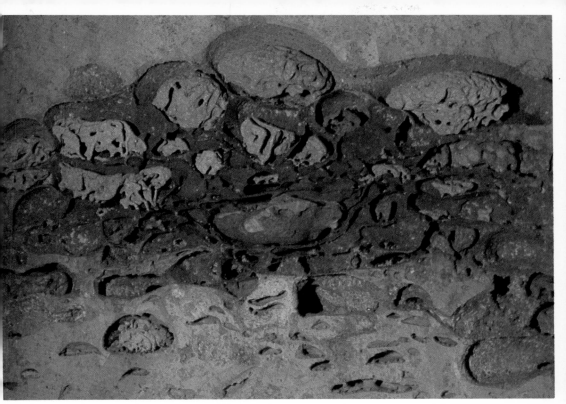

Plate 59. *Cross section through the nest of* Macrotermes carbonarius, *and, below, schematic rendition. (1) Royal cell (not opened), the center and core of the colony; (2) brood chambers; (3) a storage chamber for comminuted leaves; (4) fungus chambers. Walls consist mainly of the feces of the termites, hence their dark color. Light-colored area lower down is the nest of another species of the genus* Macrotermes *that also cultivates fungi. These termites are white and shun the light whereas* M. carbonarius *is dark and operates in daylight. Galleries lead from the nest over a distance of 1–2 cm. to the area immediately around it where the termites collect and comminute fallen leaves. Tropical rain forest, Tasek Beva Swamp, Malay Peninsula. (See pp. 132, 137.)*

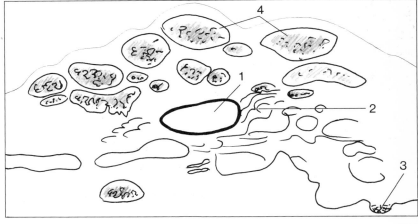

Plate 60. Termitary with chimneys (Macrotermes). *Avash National Park, Ethiopia. (See pp. 129, 143.)*

Plate 61. Termitary of Macrotermes subhyalinus (formerly Macrotermes bellicosus). Lake Manyara National Park, Tanzania, Africa. (See pp. 129, 143.)

Plate 62. Termites' mound with rain roofs on a tree buttress. (See p. 138.)

Plate 63. Mushroom-shaped termitaries of Cubitermes. The mound on left shown in transect to reveal chambers of the interior. (See p. 138.)

Plate 64. An arboreal anteater Tamandua tetradactyla *(length without tail about 55 cm.) has opened a nest of tree termites and reaches for the termites inside with its sticky tongue. (See p. 131.)*.

though they are neighbors, they have little to do with each other.

If we turn our attention once more to plate 59 (p. 133), we see in the center the cell of the royal pair (1). Life inside a royal cell is shown in figure 57 (p. 132). This illustrates how the queen, who spends her whole life in this confined space, is constantly fed and tended by troups of workers whose small size enables them to go in and out through the narrow entrance holes in the walls of the chamber. Avidly, they take over the eggs as soon as they leave the maternal abdomen (fig. 57, right) and carry them into nearby chambers. The king (2) always stays with his queen. He may reach an age of several years, like his consort, and copulates with her repeatedly.

A transect through the whole colony shows not only the royal cell and the small surrounding chambers of the nursery area where eggs and young larvae are cared for (2 in pl. 59, p. 133). It reveals also a group of storage chambers (3) where leaf particles that the termites collect and carry to the nest are kept, and other chambers, somewhat larger, where they cultivate the mycelium of a mushroom (*Termitomyces*) on a compost made from these leaf particles, or else wood particles, mixed with feces, as the leaf-cutter ants cultivate their fungus on comminuted leaves. The fungus converts the lignin of wood and other plant parts into compounds easy to assimilate. Many young larvae visit the fungus beds (4) where they find highly suitable food. The nest is closed against the outside world by its compact protective layer, but subterranean passages lead some two to four meters into the area around it where workers harvest and comminute their fallen leaves.

Not all mound-building species are as highly organized as these. *Cornitermes cumulans,* a South American species, produces mounds of considerable size, but the arrangement of their many chambers is irregular, and there is not the same degree of differentiation. There is no royal cell, so that the queen, though large, can move freely from one chamber to the next. Pierre-P. Grassé has described the gradual development of such a nest. Until it has reached a diameter of thirty to forty centimeters, the nest is egg-shaped and completely underground. A narrow air space insulates it from the loamy substrate on which its lower end rests. Galleries which lead away from this point, and later from other points as well, traverse the air space as solid tubes (fig. 58, right). As the nest grows, soil is brought up from the depth and heaped above (fig. 58, center), foreshadowing the future mound. The early soil nest is then gradually dismantled and converted into a permanent mound nest. The particular mound shown in figure 58 (left) was 1.60 meters high and 1 meter across at the base. How such a complete reconstruction comes about is a mystery. But we have even more cause for wonder when we consider the whole range of termite buildings and the way they are adapted to the most diverse climatic conditions of the countries they inhabit.

Take, for example, certain species of the genus *Cubitermes* that live in tropical rain forests. They put roofs with overhanging eaves on their tall mounds, which make them look like pagodas and which serve to keep the torrential rains off the main structure (fig. 59 and pl. 62, p. 135). Other species build the whole termitary in the shape of a mushroom (pl. 63, p. 135). The mound on the left is cut open to show the chambers. Termites living in arid zones do not build such roofs, showing that they are definitely umbrellas, not sunshades.

The treeless steppeland of Australia, baked by the scorching heat of the midday sun, is the home of the compass termites (*Amitermes meridionalis*). Their towering structures, which may be up to five meters high and three meters long, look as if they had been compressed from two sides (pls. 58 a and b, pp. 130 and 131). Their two short sides face exactly north and south, so that the surface exposed to the rays of the midday sun is small, while the long sides catch the evening and morning sun. In the cold season, the termites find their preferred temperature by congregating on the east side in the morning and on the west side in the evening. A traveler can quickly find his bearings by looking at the direction of these mounds. But how do the blind termites orient them so

Fig. 58. A termitary grows above the ground. Three stages in the development of a nest of Cornitermes cumulans: *a cross section to the left. Right and center, diameter of the nest roughly 30 cm. The mound on the left was already 160 cm. above ground and was still growing.*

perfectly without using a compass? We remember the ability of the bees to orient their combs by the aid of the earth's magnetic field (cf. p. 93 f.). The method by which the compass termites achieve their spectacular results has not yet been studied. However, laboratory experiments have shown that other termites are sensitive to magnetic forces.

Günther Becker notes that in his observation cages the sexual forms of his experimental species (*Macrotermes* and *Odontotermes*) settled either in a north–south or an east–west direction and that if their cages were turned, they would return to these positions within a few hours. The same preference for these positions was also observed in natural mounds. When termites of various species were placed in an artificial termitary in the center of a round glass dish with access to wood particles spaced all round the periphery of the dish, most of the communication galleries built by them for the purpose of collecting wood particles ran due north–south or due east–west. Why these directions were preferred was not clear. It is possible that such basic orientation in the magnetic field helps termites to keep to straight lines when building their galleries, just as the "sun compass" aids ants, bees, and other creatures of daylight in finding and keeping directions. That termites are, in fact, guided by the lines of the magnetic field was borne out by experiments in which these lines were artificially deflected as in the experiments with bees. The direction of their galleries was altered correspondingly, indicating that the termites, like the bees, use the magnetic field for orientation. It is therefore highly probable that the compass termites align their hillocks by reference to the earth's magnetic field. This does not in any way diminish our admiration for the harmonic execution of their buildings.

Fig. 59. Roofs built by termites (Cubitermes) *as a protection against rain.*

Air-conditioning in termite dwellings. The interior architecture of many termite species is even more astounding. The distribution of the various chambers according to their different purposes is evidence of a definite building plan. But the functioning of a large termitary requires not only the systematic layout of the chambers, but convenient space for the royal cell, the quarters for the different age groups, the fungus gardens, and the associated network of communications.

When a mound of *Macrotermes bellicosus* (formerly, *Macrotermes natalensis*) has reached a height of three to four meters, it contains more than two million termites. They live, they work, and they breathe. Their oxygen consumption, which has been measured, is considerable. Without ventilation they would all be suffocated within twelve hours. How is the termitary ventilated? Its solid surface shows no signs of windows.

Professor Martin Lüscher has studied the termitaries of this species on the Ivory Coast of Africa. These insects

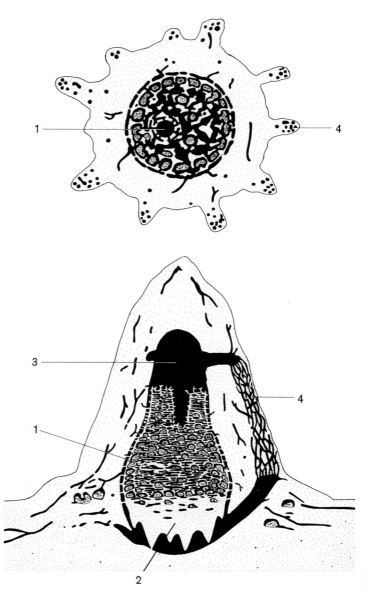

Fig. 60. Top, cross section; bottom, longitudinal section through the nest of Macrotermes bellicosus *(formerly,* natalensis) *from the Ivory Coast, Africa. Air spaces are shown in black; fungus gardens are dotted. Some of them are outside the nest proper. Both sections go through (1) the royal cell; (2) cellar; (3) air space above the nest; (4) supporting ridge, or buttress, with air ducts. Height of the mound about 3½ m.*

have established a strange and ingenious ventilation system. A cross section made vertically through the center of the mound (fig. 60, bottom) shows the nest proper, which is almost round, with its royal cell in the center (1), its many chambers and passages. Between it and the thick, hard outer wall there are narrow air spaces. Below it there is a larger air space, the "cellar" (2). The central structure rests on conical supports and is further anchored by lateral struts. Another air space above it (3) reaches a long way into the nest proper, like a chimney. On the outside of the mound, ridges or buttresses run from top to bottom (4). These are clearly shown on the horizontal cross section in figure 60 (top). Channels as thick as an arm radiate from the upper air space into the ridges (fig. 60, bottom right), where they divide into many small ducts. These come together again to form channels as wide as the first leading into the cellar. Though termites are found in all these structures, they do not act as ventilators as, for instance, bees do when they ventilate the hive by fanning with their wings. The ventilation system of the termitary is completely automatic. This is how it functions.

The air in the fungus chambers is heated by the fermentation processes taking place there. Like any tightly packed group of breathing animals, the termites themselves cause a rise in temperature. This hot air rises and is forced by the pressure of the continuous stream of hot air into the duct system of the ridges. The exterior and interior walls of these ridges are so porous that they enable a gas exchange to take place. Carbon dioxide escapes and oxygen penetrates from outside. The ridges with their system of ducts might be called the lungs of the colony. As has been experimentally confirmed, the air is cooled during its passage through the ridges; this cooler, regenerated air now flows into the cellar by way of the lower system of wide ducts. From there it returns to the nest via the surrounding air space, replacing the rising warm air.

A traveler through the Alps will soon become aware that the traditional houses—those that have not been blighted by the leveling process of our mass civilization— have their own charming local peculiarities. It may be the characteristic shape of the roof line, an original type of carved ornamentation, or some other special feature. Termite mounds can have similar local differences. The same species whose ventilation system has just been described

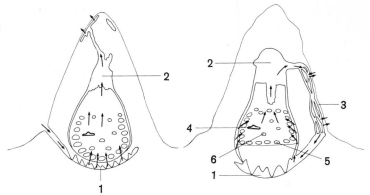

Fig. 61. Diagram of the circulation of air in nests of Macrotermes bellicosus (natalensis): *right, in a nest from the Ivory Coast; left, in a nest from Uganda. The arrows indicate the direction in which the air moves. (1) Cellar; (2) upper air space; (3) ducts in the ridges; (4) royal cell; (5) fungus chambers; (6) brood chambers.*

from West Africa is also found in Uganda. There is not the slightest morphological difference to suggest that the termites from the two countries might belong to different varieties. But their ventilation systems are not the same.

The Uganda mounds of *Macrotermes bellicosus* (formerly *natalensis*) lack the conspicuous ridges with their air ducts. Instead, the warm air rising in the nest is led by ducts from the upper air space into flat chambers under the dome (fig. 61, left) whose walls are so porous that the air can escape to the outside. The cellars of their mounds are open to the outer air by means of wide channels, but they are closed off from the nest and inaccessible to the inhabitants. However, the floor of the nest wall is so porous that it allows the fresh air from the cellar to penetrate. In this way a continuous stream of fresh air through the nest is maintained. The two systems are compared diagrammatically in figure 61. Both methods, internal circulation of regenerated air and renewal of air from outside, appear to function equally well. How these different solutions of the same problem developed in the course of the evolutionary history of the species is not known.

Other termitaries (those of *Macrotermes* and *Odontotermes* species) have conspicuous tubes, like chimneys, whose function has not been studied in detail but which clearly provide ventilation. These chimneys are open at the top and are connected with ventilation shafts reaching through the nest proper into the ground below where they are closed at the end. Normally, this cavity is completely separated from the nest proper by a thin wall which, presumably, is permeable to air. No termites are found in them except during periods of construction or repair. The chimneys often collapse during heavy downpours. At such times many termites quickly collect and start repairing (according to a personal communication by Professor Martin Lüscher). Plate 60 (p. 134) shows a particularly impressive example of long chimneys. I am indebted for this photograph to Heinz Sielmann, who took it in 1972 in Ethiopia. The arrangement of the chimneys suggests that the builders may have been *Macrotermes subhyalinus* (*bellicosus*) like those of the termitary shown in plate 61 (p. 135). With these termites, too, there is considerable variety between the nests of different localities. However, the exact species has not been ascertained.

In general, termites of a given area keep strictly to their traditional building style. This makes it even more remarkable that meaningful reactions to extraordinary situations, or what one might call emergencies, have been observed. When a termite mound was enveloped in a plastic tent so that ventilation was seriously impeded, the termites managed within forty-eight hours to build new structures at the top of the mound, which looked somewhat like small pointed hats and had exceptionally porous walls so that they functioned as a new ventilation system.

One more example of the great variety of solutions to the ventilation problem. *Apicotermes gurgulifex* builds an oval nest, about twenty centimeters high (fig. 62). It is embedded in the soil but is insulated from the surrounding earth by a mantle of air. In the interior of the nest there are flat chambers joined by a central spiral path. Its outer wall shows a pattern of raised, ring-shaped configurations. Each ring surrounds a slit leading into circular passages inside the exterior wall of the nest which, in turn, are connected with the rooms inside, so that all pores contribute to the ventilation of all living quarters. One cannot but marvel at the workmanship of these structures whose openings are spaced and shaped as regularly as if they were punched by a machine.

Ventilation is not the only problem of termite communities. Water is another. A great deal of water is needed because the inhabitants with their tender skins require a humid atmosphere. In the nests of *Macrotermes* described above, relative humidity is eighty-nine to ninety-nine per cent. Much water is also needed for consumption, for making mortar, and for other purposes. In arid regions, termites may dig to enormous depths to tap the ground-water table. Figure 63 shows diagrammatically the water system of *Trinervitermes*, a species that builds small mounds, twenty-five to thirty centimeters high, in

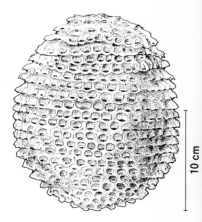

10 cm

Fig. 62. Nest of a termite species (Apicotermes gurgulifex) that uses its own excrement to fashion a harmonious structure. The nest, about 20 cm. high, lies below ground and is surrounded by an air space. The surface is pierced by ventilation slits, each slit being surrounded by a raised ring. So precise is their pattern and spacing, slits appear to have been made by mechanical puncher.

143

the African savanna. The center of the nest contains the breeding and living areas; the outer chambers are used for stores which are brought in through lateral galleries. Several vertical shafts lead into the ground, including one which penetrates to great depths. Some desert termites were found that drive bore holes down to water at a depth of some forty meters. The construction of such deep shafts through loose soil is a truly prodigious feat of civil engineering for these small animals.

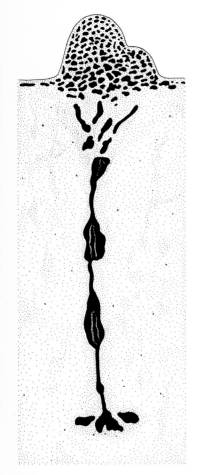

Fig. 63. Diagrammatic cross section through the nest of a termite from an arid region of Africa (Trinervitermes). A deep shaft leads from the mound to the level of the ground water.

Building techniques. Termites that do not excavate their nests from soil or wood use a variety of building materials.

The disposal of excreta is a problem that has to be solved in all human settlements. Long before human beings were faced with it, termites had found a solution that is both simple and practical. Many species use their feces for the building of their own homes. Admittedly, this solution can only be used for excreta that dry quickly and do not putrefy. The *Apicotermes* species whose ventilation slits have already been described (fig. 62, p. 143) fashion their nests in this manner, turning their feces into a work of art. We do not know how they do it. All that can be seen is that the termites deposit a small drop of excrement where it is needed for the construction of the nest, then turn around and spread the mass while it is still soft, and smooth it with their mandibles.

However, most species employ pellets of soil, grains of sand, particles of wood, and other extraneous matter as their main building material and use their excreta merely as a binder. They may also use a hardening saliva for this purpose, or a combination of both.

Apart from solid nests, several types of which have been discussed, termites also fashion more fragile structures that not only resemble wasps' nests (fig. 64 a) but are made in a similar manner. Three times the inventive genius of the social insects has discovered the technology of papermaking and its use for building. The termites use paper pulp made of masticated wood mixed with saliva or excrement not only for the outer walls of their dwellings but also for the construction of their living, breeding, and storage chambers, including a royal cell as the centerpiece (fig. 64 b). Many different species produce such carton nests.

We biologists would dearly like to know how and in what manner the activities of various workers of a termite community are coordinated. How is it possible for these small animals to make such harmonious and ingenious structures? Unfortunately, it is not possible to watch them undisturbed at work in their nests. Attempts have therefore been made to observe their behavior in other ways.

The subterranean passages and covered surface runways by which termites reach their collecting areas and other places from their nests have already been mentioned. Many years ago, Karl Escherich had the idea that it ought to be easier to observe termite building operations on such runways than on the actual nests. In Ceylon

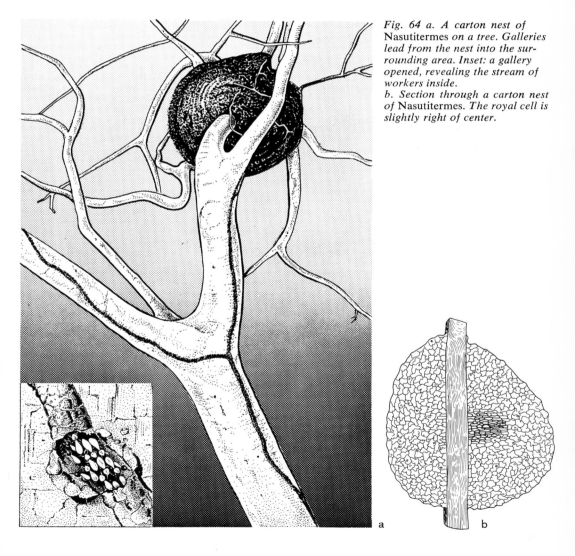

Fig. 64 a. A carton nest of Nasutitermes *on a tree. Galleries lead from the nest into the surrounding area. Inset: a gallery opened, revealing the stream of workers inside.*
b. Section through a carton nest of Nasutitermes. *The royal cell is slightly right of center.*

a   b

he found a subterranean nest of *Eutermes ceylonicus* from which a covered runway led up the stem of a coconut palm into its crown. He removed a small portion of the roof of the tunnel in order to see how it would be repaired. He did not actually see the view shown diagrammatically in figure 64 a (inset). After he had damaged the tunnel, he just saw a few termites disappear in it. Then all was quiet for a time. After a pause of several minutes, one nasute soldier cautiously emerged from the tunnel, inspected the extent of the damage with great care, and withdrew. Soon several soldiers appeared and took up positions at the top and bottom end of the opening. Only their pointed noses and their wavering antennae were visible. More soldiers lined up along both sides of the damaged section. Next, a group of workers appeared and began to repair the damage starting at both ends, but he could not see much of them. Only now and then the tip of an abdomen became visible between two soldiers as a worker deposited a large drop of excrement on the edge of the broken tunnel, and soon after a head pressing a small soil particle into the excrement. Systematically, building brick by building brick, the damage was repaired in a few hours. In this case, the direction of working was given by the old line of the tunnel and by the line of sentinels. Other research has also suggested that soldiers exert some influence on building operations. But this is still conjectural and certainly not universal.

Recent research into gallery-building has not provided much new insight into the organization of labor. Apparently a scent trail, laid by termites that have gone this way before, frequently indicates the line of a new gallery.

Road-building termites do not always cover their roadways. *Odontotermes magdalenae* only pave them with soil particles moistened with saliva. *Trinervitermes* use hardening droplets of excrement as paving stones.

It is probable that in work of this nature the builders are guided by scent trails on the ground. Such an explanation is not applicable to those galleries or tubes that termites build into the air toward a specific target. Over short distances, however, one may assume scent as the guide. I am indebted to Professor Martin Lüscher for the interesting photograph shown in plate 65, and for his permission to publish his personal communication describing the situation:

"A group of about 200 *Reticulitermes* was kept in my laboratory in the glass container seen on the right-hand side, which served as an artificial nest. A bottle with a cork stopper happened to stand nearby. One night the termites gnawed through the cork cover of their container and built a set of aerial tubes. With one of the tubes, they found the bottle cork and started to attack it at once. I found this unplanned 'experiment' fascinating and decided to photograph it immediately. I believe it proves that termites can smell a cork over a distance of five centimeters and that they orient their galleries by smell."

It may be that planned experiments will some day throw light on the process of gallery-building. But the major problems of termite architecture concern the marvelous structure of their mounds, and so far very little has been discovered about the way these are built. In figure 65 is shown a diagrammatic sketch of a worker engaged in building a wall. First, the animal examines the place where it is working, then it turns around and deposits a drop of excrement on the place just investigated. Next, it picks up a pellet of soil in its mandibles, turns around a second time, and presses it into the hard-

Plate 65. Galleries built through the air. The glass container on the right harbored a colony of Reticulitermes. *Overnight, the termites gnawed through the cork, and after several abortive attempts, built a complete gallery to a neighboring container, attacking its cork stopper. Observation by Dr. Martin Lüscher, Bern, unpublished.*

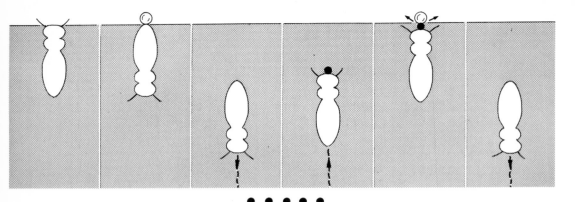

• • • • •

Fig. 65. Diagram for the sequence
of operations when soil particles
are built into droplets of excrement.

ening excrement. After testing whether it is firmly embedded, the worker repeats the same process with the
next building brick.

In figure 66, a group of workers is constructing an
arch. One of them has just released a drop of excrement
from the anus (white arrow); others put a particle of
soil or sand into the excremental mortar. The sequence
of operations may be slightly varied, saliva may be used
instead of excrement and so forth, but each worker works
always in a tiny area of the whole structure, mechanically
adding particle after particle. A worker moving to another
part of the building will continue whatever work is required by the actual state of building operations in that
place.

Fig. 66. Macrotermes natalensis
(formerly M. bellicosus) building an
arch, the two halves of which are
about to be joined. The building
material consists of droplets of
excrement and soil particles. The
arrow to the left points to a droplet of excrement which has just
been deposited. The workers on
the ends of the arches put down
soil particles.

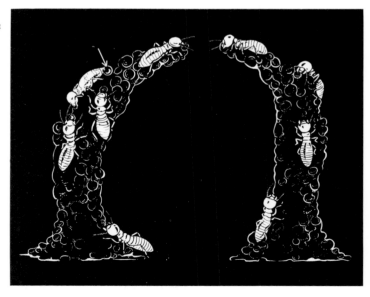

If we imagined for a moment that termites were as tall as human beings, their tallest hillocks, enlarged on the same scale, would be nearly a mile high, four times the height of New York's Empire State Building. How can a planned construction of such mighty buildings be brought about? Or the ingenious ventilation systems (figs. 60, 61, pp. 140 and 142)? Or the meticulously modeled exterior skin of the small nest of *Apicotermes* (fig. 62, p. 143) with its ventilation slits and the spiral staircase inside it? Where is the architect?

A question that will be asked here is whether termites can communicate with each other. There is no doubt that they can and do. They can mark a trail with scent so that other termites can follow it, and they can give knocking signals by striking a hard surface with their heads. But the information content of both modes of communication is small. The scent trail may lead to a goal, but it cannot explain what should be done there. Drumming is an alarm signal by which soldiers or workers induce other workers to flee into the interior of the nest. The vibrations produced are perceived by other termites with the aid of highly sensitive tactile organs in their legs. But it is just a general warning signal.

The brain of insects is housed in the head just as it is in human beings. Though its structure differs from that of vertebrates, the insect brain similarly exhibits an increase in complexity and in number of nerve cells among higher-developed species as compared with more primitive ones. It has long been assumed that a certain part of the brain of social insects, the "mushroom body" (*corpora pedunculata*), serves the higher mental functions, especially the formation of associations. This is where the data supplied by the sense organs are coordinated and the insects' actions are determined. Recently, electrophysiological experiments have confirmed this hypothesis and made it possible to elucidate further the significance of the "mushroom body." This organ is remarkably well developed in bees, wasps, and ants, and the assumption is justified that it is connected with the ability of insects to learn and with the complexity of their behavior. However, according to studies by P. E. Howse and Williams, that particular part of the brain is not well developed in termites. Nor do the more highly developed species that erect those remarkably complex structures possess more nerve cells than the primitive forms that build simple nests.

Hence, it is probable that individual experience enters less into the building operations of termites than into those of bees and other Hymenoptera. And yet their finished structures seem evidence of a master plan which controls the activities of the builders and is based on the requirements of the community. How this can come to pass within the enormous complex of millions of blind workers is something we do not know. One can try circumlocution with learned words, but I think it is better to say, quite simply, we do not understand. Here, as so often in the science of life, the investigating human spirit must bow before the unknown.

Vertebrates

Vertebrates are identified by zoologists as a subphylum of Chordata, the classification of animals that develop notochords at some stage of their lives. The vertebrates include five large well-known classes: fishes, amphibia, reptiles, birds, and mammals. Some forty-two thousand living species of vertebrates have been described, compared with over eight hundred thousand species of arthropods of which the majority—seven hundred and fifty thousand—are insects. Hence, it is hardly surprising that the varied building activities of arthropods have offered a greater field of research to inquiring biologists and have gained their special interest by this amazing variety.

On the other hand, we feel a much greater affinity with vertebrates than with insects, because we too, by reason of our anatomy and general behavior, are classed among the vertebrates. Animals with a spinal, or vertebral, column developed as a separate phylum independently of the arthropods and have solved many biological problems in an entirely different manner. They alone achieved that development of the brain from which man's mental powers and culture could result.

Humans can apply their powers of thinking and reasoning to create buildings that meet their tastes and requirements. By contrast, insects and spiders, as we have seen, can build highly complex structures admirably adapted to their needs without previous practice or experience, solely on the basis of their innate instincts. Observing the achievements of vertebrates in this field, we shall find that as regards skill of construction the nests of many birds are comparable to the works of insects. But there the comparison ends. The bodies of insects and vertebrates are totally different. If we "look a bee in the eye," look into her miraculous compound organs of vision with their ten thousand facets pointing in all directions, we do not experience the same emotional response as when we look into the eye of a bird. With the bird we feel the eye to be the mirror of a *mind,* like the human eye. Therefore, many people are inclined to believe that birds approach their nest-building with reason and foresight.

But, in fact, birds' activities are also ruled by instinct, even if learning and individual experience do play a role in guiding their actions. Few species of the lower vertebrates, that is, the fishes, amphibia, and reptiles, build nests of any kind, and no evidence suggests that these species are guided by any form of reasoning.

FISHES

Salmon as modest builders. Few people could name more than a handful of the twenty thousand species of fish that have been identified, though they may know the names of some edible species, such as trout and salmon. Salmon are not only good to eat and much sought after by anglers, they are also justly famous for their migrations. A year or two after their birth in a mountain stream they migrate down to the sea and there disperse to places hundreds, and often thousands, of miles away. Several years later they return, well fed and strong, to their native river basins and, migrating upstream, find their way back to the very waters where they hatched. They are guided in this amazing feat by the peculiar smells of the waters they passed through on their downstream journey in early youth which have been indelibly imprinted on their memory.

We cannot concern ourselves here, however, with the fascinating matter of their orientation. When the salmon arrive at their destination, they, or more accurately, the females, choose a suitable spawning site, preferably in a stream with a swift current and a gravel bed, and embark on building activities, admittedly simple. With powerful strokes of their tails and hind parts, they hollow out a shallow pit, ten to twenty centimeters deep and one to two meters long, in the direction of the current. The males watch, but do not assist. Then they come close and, after some courtship display, eject their semen into the pit as the females lay their eggs in it. To make sure that their eggs are not washed away by the current and hidden from sight, the females cover them with gravel with a few more strokes of their tails. Beyond this, salmon do not tend their progeny, and such parental care as they exercise must be called primitive. And yet the shelter of the gravel nest provides sufficient protection for their eggs and for the newly hatched brood during the period in which the little fish still live on the rich yolk of their egg-sacs; in this way, the species has been able to maintain its num-

Plate 66. Male paradise fish below his bubble nest. In background, the female. (See p. 158.)

Plate 68 a. A male stickleback is building a nest. (See p. 161.)

Plate 67 (below). Fighting fishes. The pair below the bubble nest. The female nudges male to indicate her willingness. (See p. 159.)

Plate 68 b. The stickleback swims over his nest and glues its component particles together with a secretion from his kidneys. (See p. 162.)

Plate 69. The leatherback turtle (Dermochelys coriacea) *lays her eggs on a tropical sandy beach. (See p. 170.)*

Plate 70 (above). *A young brush turkey has hatched and has worked its way from the inside of mound to the surface. It runs away immediately, completely self-reliant. It never meets its parents.*

Plate 71 (left). *The male brush turkey builds his mound in the Australian forest. When he had finished, this mound was 4 m. wide and 1½ m. high—not a mean performance for a bird no bigger than a largish domestic fowl. (See p. 177.)*

Plate 72. *Thermometer fowl* (Leipoa ocellata) *protect the small amount of compost collected in the arid open bush of Australia by heaping on it a high mound of sand. The male has just dug down to the depth of the eggs and is testing the temperature with his beak. The female waits at the top, ready to lay another egg into the pit as soon as both partners find the temperature suitable. (See p. 179.)*

Plate 73. *A nest of the little ringed plover. Because of their coloration, eggs are excellently camouflaged in their pebble-strewn hollow. (See p. 184.)*

Plate 74 a. *An eider duck brooding.*

Plate 74 b. *Nest of an eider duck cushioned with her down feathers. (See p. 185.)*

bers over thousands of years, producing about fifteen thousand eggs per female each spawning season. (Carp, which drop their eggs in unprotected places, need to produce about half a million to one million eggs to obtain the same result, and codfish about nine million.) Only in the last hundred years have the numbers of salmon declined sharply in many areas, as a result of urbanization and river pollution. Nature is powerless against the rapid, far-reaching, and often senseless changes in the environment brought about by man.

Bubble nests of labyrinth fishes. The nurseries of salmon in the gravel beds of mountain streams are hard and stony. They form the greatest contrast imaginable to the bubble nests created by the labyrinth fishes on the surface of still waters as shelters for their broods. This interesting group lives in remote regions of the tropics and subtropics of Asia, rather a long way from the homes of most aquarium keepers. In spite of this distance, the labyrinth fish *Macropodus opercularis* (paradise fish) was introduced from Indochina into Europe as the first tropical aquarium fish more than a hundred years ago. This brightly colored creature soon became a favorite of aquarium owners because it is comparatively easy to keep and can be induced to build a nest and propagate in small tanks, provided the temperature is right. Since the advent of air transport has allowed the importation of more delicate animals, its popularity has declined somewhat in favor of other brightly colored tropical species, such as the closely related fighting fishes.

The name "labyrinth fish" refers to a peculiarity of the family closely connected to its nest-building activities. All fish possess interior gills under their gill covers, which contain gill laminae that take up oxygen from the water. Labyrinth fish possess an additional air-filled breathing organ in the upper part of the gill cavity. This space is divided into labyrinthine chambers by protruding bone formations covered with mucous membranes rich in blood vessels, thus providing a large surface that can absorb oxygen directly from the air. By frequently snapping for air, the fish constantly renew the air of the cavity when it gets stale. Because of this supplementary method of respiration, they often live in waters extremely poor in oxygen and thrive in them. With the approach of spawning time, each male seeks a suitable place for a nest on the surface of his water. In contrast to salmon, it is the male who is the active partner. His colors become brighter, and when he is excited he can make them truly radiant. He now snaps for air far more often than he would need to for breathing. Releasing the air under water, he causes a number of air bubbles to collect on the surface. Eventually these form a ball of foam which becomes attached to a floating leaf or to the stalk of an aquatic plant (fig. 67 and pl. 66, p. 153). At other times the bubbles formed when the fish snaps for air and ejects stale air would burst as rapidly as air bubbles normally do on reaching the surface, but during this period the bubbles are unusually stable. Through the effect of a hormone, the number of mucus cells in the mucous membranes and the amount of their secretion increase considerably before the onset of the spawning period; hence the mucus that mixes with air in the fish's mouth cavity imparts considerable stability to the bubbles.

The females are interested in these activities, and from time to time one of their number will venture close. But the owner of the territory is in such an aggressive mood that he chases even females away. Once the nest is completed, however, his aversion gradually diminishes. Eventually the intruding female is tolerated and mating takes place below the nest. The male "embraces" the female from below by bending his body to a U-shape so that his mouth and tail both touch her back, and both partners

Fig. 67. Paradise fish (Macropodus opercularis) *making a bubble nest.*

158

eject their sexual products simultaneously. Because the yolks contain a large globule of oil, the eggs are lighter than water and rise automatically into the bubble nest. If any should drift away, the male catches them with his mouth and spits them into the nest. This mating behavior may be repeated several times. Once she has finished spawning, the female is no longer allowed near the nest. The male alone guards it, maintains it, and replenishes the foam. The father guards the fry for the first few days after hatching. If they roam too far, he picks the strays up in his mouth and spits them back into the nest. However, his protective urge is of short duration. Soon the young fishes disperse and the foam disintegrates.

Aquarium enthusiasts delight in paradise fishes because of their beautiful coloration during the period of territorial defense and mating. But this species is surpassed in aggressiveness and fierceness toward rivals by another group belonging to the labyrinth family, the previously mentioned fighting fishes. One of the most beautiful fighting fishes, *Betta splendens,* has become widely popular as an ornamental aquarium species. It is one of a dozen or so Siamese and Indonesian representatives of this group, most of which build bubble nests (pl. 67, p. 153). These fishes develop colors of magical splendor during fights with rivals, and it is customary in Thailand to enclose two males in a small tank to make them fight. In these circumstances, the weaker fish cannot escape by flight as he would under natural conditions. Each owner bets large sums that his fish will overcome and kill the other, and spectators gather to enjoy the fishes' beautiful colors and their exciting battle to the death. People are the same the world over, whether in regard to fighting fishes or fighting cocks or bulls pitted against matadors in the arena. Indeed, the more highly developed the creatures that can be made to savage each other, the greater the humans' enthusiasm for the cruel spectacle.

The refuge of the sand goby. The sand goby (*Gobius minutus*) builds its nest in an entirely different manner. These fish dwell at the bottom of their waters like other Gobiidae. They mostly rest on their pelvic and pectoral fins, or else swim jerkily about. They are lively little creatures who constantly watch their surroundings with large eyes.

Their nests were discovered along the French coast. The male, who in this species, too, is the nest-builder,

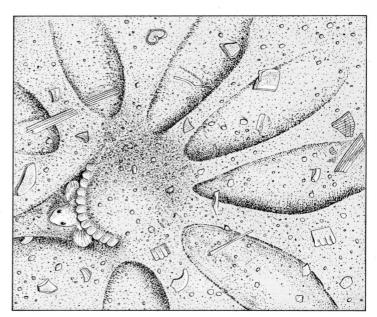

Fig. 68. Sand goby (Gobius minutus) *under its sand-covered shell. A small ditch leads to entrance of the nest. Depressions radiating from shell indicate where fish removed sand which he has heaped onto the shell with his pectoral fins. Only the rim of shell remains visible.*

searches for the empty shell of a bivalve, preferably one similar to the heart-shaped shell of *Cardium edule,* the common cockle. If he finds this lying with the concave side up, the fish will seize it with his mouth and turn it over. He then takes some of the sand underneath the shell into his mouth and carries it away. Next, lying on the ground, he heaps sand on top of the shell with his breast fins, leaving only a narrow channel for access, which he reinforces with the sticky mucus that covers his skin (fig. 68). He is now ready to conduct a female to the nest entrance. She will slip in, turn on her back, and deposit her eggs onto the ceiling of the nest where they will adhere. The male will guard the eggs until the young fish emerge. These soon disperse and, at first, live in open water.

Males of the paradise fish, fighting fish, and gobies deserve credit for the way they guard and defend their nests. But the building activities of these fish are rather primitive. Others can do better, as the sticklebacks demonstrate. They owe their name to the sharp spines with which they are equipped. Several of these are located at the animals' back in front of their dorsal fins, and one on each side, at the base of the pectoral fins. The fish can raise the dorsal spines and make the lateral ones stand away from the body; and they can fix them rigidly in these positions by means of special locking joints. They thus present a formidable sight, and even fairly large predaceous fish hesitate to swallow such sharp-spiked prey.

Nests of sticklebacks and wrasses. The fresh waters, brackish waters, and coastal waters of the northern hemisphere contain several species of stickleback. Their fame dates from a lecture at the French Academy of Science in 1844 when the learned assembly heard with great surprise of fishes that build nests in the water, the construction of which resembles certain bird nests and into which the female fishes lay their eggs.

I shall take as my example the three-spined stickleback (*Gasterosteus aculeatus*), a species widely distributed in North America and Europe. When the time of spawning approaches, shoals of these fishes move from deeper to shallower waters. The males, resplendent in nuptial raiment with brilliant red throats and bellies, separate from the swarm, and each searches for a suitable site, preferably one with a shallow bottom of sand and gravel furnished with aquatic plants. They vigorously defend their chosen territory in the manner of all territorial animals. Nest-building is started with the excavation of a shallow pit. Then follows a collection of plant material which the fish either picks up from the sea bottom, or bites off from nearby vegetation (fig. 69 and pl. 68 a, p. 153). It largely

Fig. 69. A stickleback building his nest.

consists of strands of algae, leaves of aquatic plants, parts of roots and other pieces. But it is not merely brought together in a heap. With slow movements, the stickleback swims repeatedly across the loose pile while he secretes a sticky substance from his kidneys that serves to glue the various pieces together (pl. 68 b). The greater cohesion which has thus been imparted to the walnut-sized little ball prevents it from disintegrating when he next proceeds to bore an opening into it with his snout. A hole made on the opposite side completes the tunnel; the interior cavity is enlarged by pushing up the ceiling. Once the nest is ready, the stickleback usually has no difficulty in finding a willing female. Niko Tinbergen has shown in great detail how the male stickleback leads the female to the nest entrance with strikingly peculiar swimming movements, how he induces her to enter it by a pointing movement with his snout, and how finally, by prodding her tail as it protrudes from the nest, he induces her to spawn. We have here an excellent example of the way in which a specific behavior of one partner releases a corresponding reaction in the other; the whole sequence of instinctive actions is built up of "sign stimuli" emanating alternately from the female and the male. In the final act, the female slips out of the second hole and the male enters by the first hole to fertilize the eggs. Other females may later be induced to spawn in the same nest until it contains several hundred eggs. Then the male alone remains with the nest and cares for the brood in a manner that, so far, we have not yet encountered in any other fish. He chases away all rivals, all females, and, indeed, any other creatures that approach too close. He takes up a position in front of the entrance and fans fresh water onto the eggs in the nest by movements of his front fins. Should there be a shortage of oxygen in the nest cavity, he makes further openings in the nest walls to ensure adequate ventilation. If the nest is damaged he repairs it. The young hatch after about a week and soon leave the nest though they stay close to the nest in a compact little swarm, guarded by their father. Any stragglers are picked up by him with his mouth and returned to the fold. After a few weeks his guarding instinct wanes, and the young sticklebacks, collecting in larger shoals, go their own way.

The sticklebacks are the best-known builders among the fishes that construct nests in the way birds do, but they are not the only ones. Another is *Crenilabrus ocellatus,* a Mediterranean member of the family of Labridae

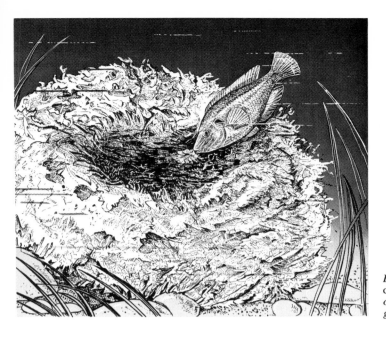

Fig. 70. The wrasse Crenilabrus ocellatus *builds his cup-shaped nest on the sea floor from algal strands growing in the vicinity.*

("Lip-fishes," or wrasses) whose many species are wide-spread over the seas.* Somewhere close to the shore, the male, using strands of green algae, builds a cup-shaped nest resembling in its form that of a blackbird. The fish collects the algae in the vicinity, carries them in his mouth in batches to the nest site, and pushes each batch into the mass already assembled so as to turn the whole into a firm heap (fig. 70). The female attaches her eggs singly to the algal strands, and the male fertilizes them immediately. The male then covers up the first batch of eggs with more algae, and the female lays more eggs. Eventually the eggs are distributed in the bowl-shaped algal mass like raisins in a pudding. The male guards and defends his nest for a considerable time. This sentinel duty is highly necessary, because a great many fishes, including those of his own species, prey on the spawn.

I do not wish to give the impression that in the world of fishes the females do not build or guard nests. Certain cichlids, fresh-water fishes from South America and Africa that are very popular with aquarium owners, excavate shallow pits for their broods at the bottom. This nest-building and the defense of the eggs and young fry

---

* For American readers, the most familiar members of this family are the tautog, or blackfish (*Tautoga onitis*) and the cunner, or bergall (*Tautogolabrus adspersus*).

Fig. 71. The "well-digger" jawfish in his tube, which is lined with pebbles and the shells of snails and bivalves.

are in some species done by the females, in others by the males, and in others again by both sexes.

Jawfishes. Jawfishes (Opistognathidae), which live in tropical seas, are related to the cichlids. They sometimes do more than provide temporary shelter for their off-spring: they build proper homes. One species inhabiting the shallow waters off the southern and southeastern coasts of Asia excavates tube-shaped dwellings in the sand or silt of the sea bottom which may attain a depth of one meter. Their anatomy fits them admirably for work of this nature. Not only do they have very large mouths, they have greatly elongated jaw bones as well, which enable them to open their mouths to a considerable width. Incidentally, their scientific name is also a mouthful: *Gnathypops rosenbergi*—their common name in German is "well-digger," which tells us more and is easier to re-member. During its digging operations, the well-digger takes great mouthfuls of soil from the bottom and empties it outside the cavity, like a live dredging machine. The lower part of the shaft is widened into a cavity. Stones and small bits of shells or corals are pressed into the upper parts of the wall, giving the entrance the appear-ance of a well lined with masonry (fig. 71).

Safely ensconced in its dwelling, the well-digger, whose large mouth is no less useful for snapping than for hole-digging, lies in wait for prey. When it feels safe, it will put its head out, then cautiously emerge to wait, motion-less, for an edible morsel to come its way. But at the first suspicion of trouble, it will retreat into the shelter of its tube, tail first; and when real danger threatens, it will dive in. This fish resolutely defends its dwelling against all comers, including members of its own species.

Mouthbrooders. Though it lies a little outside our main theme, it should perhaps be mentioned here that certain species of jawfishes * and many species of cichlids have discovered another use for their large mouth cavities. They are mouthbrooders. The males, and in some species the females, take the newly laid eggs into their mouths, guarding them (and later their fry) from danger. This behavior compels them to fast for several weeks, but the urge to protect their offspring conquers their desire to

---

* Among these can be included the jawfishes native to coasts of the southeastern United States.

swallow the eggs or the brood, something which other fishes will do without compunction if they have the opportunity. This is an interesting and successful solution of the problem of finding a sheltered home for one's offspring.

AMPHIBIA

The name Amphibia, derived from the Greek words for "double-sided" and "life," indicates the "double nature" of these animals. In the course of their phylogenetic evolution, they have changed from an aquatic to a terrestrial way of life without completely severing their connection with water. They usually lay their eggs in ponds or streams and the soft-skinned larvae grow and develop in water. Not until a salamander larva has become a salamander, or a tadpole has turned into a frog, do the gills disappear, after lungs have developed to take over the function of breathing.* By that time their skin, too, has become tough so that life on land is possible.

Most amphibia leave their offspring to fend for themselves. They lay a great many eggs and take little notice of them afterward. In a few species the brood is watched over by the male or the female, but protective devices are rare. A few examples will suffice to describe their simple structures.

Foam nests. Foam nests are comparatively frequent with frogs. They are chiefly encountered in the tropics, as with the African tree frogs (genus *Chiromantis*), the Javanese flying frog (*Rhacophorus reinwardti*), and the Omai "rowing" frog (*Rhacophorus omaimontis,* named after his home, a mountain in western China). It is highly probable that different species living in widely separated regions have "invented" foam nests independently. It is quite certain that the development of this nest-building habit had nothing to do with the bubble nests of labyrinth fishes or cuckoo spit insects. Frogs, labyrinth fishes, and cuckoo spit insects cannot have inherited this technique from common ancestors, nor can they have learned it from each other by imitation. Moreover, they all produce their foam in a different manner: the cuckoo spit larvae inject the stale air exhaled by the tracheae in the form of

* One family of salamanders (Plethodontidae) never develop lungs and breath entirely through the skin.

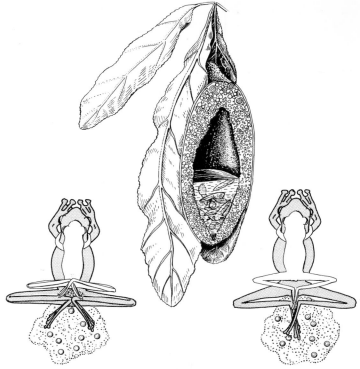

Fig. 72. Top, the Javanese flying frog. Right center, foam nest, four days after spawning. The foam is partly liquefied, and has created a mini-aquarium for the tadpoles. Bottom, couple during spawning. The male is shown in white.

little bubbles into a viscous secretion and so build up a foam cover around their bodies (cf p. 49 and pl. 25, p. 40); the labyrinth fishes take in air at the water's surface and release it again under water after mixing it with mucus to ensure the longevity of their air bubbles. And the frogs?

Their method comes closest to the method we use when we whip cream or produce a soapy lather. When the Javanese flying frogs mate, the male mounts the female and clasps her with his front legs as frogs usually do when they copulate (fig. 72). The couple chooses a good-sized leaf or settles between some smaller ones before the female lays her eggs, one by one, and the male fertilizes them on the spot. The eggs are accompanied by a mucus secretion. Both animals hold their hind legs in a bent position over their backs, and when an egg appears, they dip their feet into the mucus and quickly bring them together with rapid, kicking movements (fig. 72, bottom). Then there is a short rest until the next egg emerges. Within the space of half an hour to an hour, they have a clutch of sixty to ninety eggs surrounded by a ball of foam five to seven centimeters in diameter. The female now presses the leaves onto this ball which quickly hardens at the surface and sticks to the leaves. Having done this, the parents care no more for their brood. During the period of embryonic development, the foam partly liquefies so that the tadpoles on hatching find themselves in a sheltered mini-aquarium (fig. 72, center). There they live comfortably, drawing on the rich nutrient reserves of their yolk-sacs, until the next tropical rainstorm dissolves the outer layer of their nest and rudely washes them down into the open water.

The female of the African tree frog also builds foam nests, choosing as her spawning site a leafy shrub close to a pond. She does not protect the nest with a cover of leaves in the way of the Javanese species, but tends her brood in a different manner. She guards the nest, and she herself sees to it that it does not dry up. If the mother is caught and taken away, the surface foam of the nest soon dries and becomes so hard-baked in the tropical sun that the tadpoles cannot get out and must perish miserably. But normally the female frog ensures the necessary moisture by climbing down into the pond at intervals, where she absorbs water through her skin, then returning to the nest and watering it with her urine. As these frogs usually build their nests on leaves or branches overhanging a pond, the tadpoles just drop into the water when the foam finally disintegrates and continue their development in the pond.

Several other flying frogs behave in a similar manner. To avoid any misunderstanding, it should be explained that these frogs cannot "fly" as birds do. On taking off, they extend their long webbed toes, thus turning their feet into little kites or parachutes, and this enables them to glide over distances of several meters to another branch or down to the ground.

A tree frog builds with clay. A tree frog *Hyla faber* that builds a clay wall to make a crater-shaped nest has a much better claim to the name of "builder" (*faber*) than the makers of foam nests. The Brazilian name for this frog, which is indigenous to Argentina and Brazil, means "smith," because its croaking sounds exactly as if someone were hammering a copper plate with slow, rhythmical beats. These close relations of our tree frogs normally live in the crowns of trees, but seek shallow waters at spawning time. In calm waters, the male frog builds

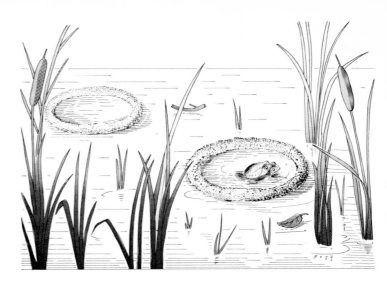

Fig. 73. The Brazilian tree frog Hyla faber *builds a clay wall to form crater-shaped clay nest in shallow water. In crater to right, a couple during spawning.*

a circular wall of clay with a diameter of about thirty centimeters (fig. 73) which, when finished, rises about ten centimeters above the water. He really works like a proper little artisan, picking up one lump of clay after another with his hands and adding them to the parapet of the crater, which he carefully smoothes, especially on the inner side. He uses his big hands, with their spatulate final toe joints, exactly as a mason uses his trowel. This work may take him a couple of days or more. When it is completed, the frog sits down in the center of the crater and calls for a mate with his powerful voice, especially at night. As soon as a female arrives, spawning and fertilization take place inside the ring wall. These frogs do not exercise any further parental care. The tadpoles are quite well protected in their nursery from the predations of fishes and other aquatic animals, though they are not too well supplied with nourishment. However, in early youth the shelter of the wall is more important.

A strange nursery for little frogs. I cannot resist the temptation of another slight deviation from my major theme in order to draw attention to an astonishing parallel between frogs and fishes. In Chile there occurs the "opossum frog," or Darwin frog (*Rhinoderma darwini*) which, as its name indicates, was discovered by Charles Darwin during his journey around the world. The female lays about thirty very large eggs with big yolks. Two or three males stay with the eggs and guard them with great patience. It takes about two or three weeks until the embryos are sufficiently developed for their movements to become visible through the egg membrane, and only then does each frog pick up a number of eggs with his tongue. Within a few days, all eggs are collected. The amazing thing is that the eggs are neither swallowed into the frog's stomach nor held in the mouth cavity, but that they are pressed down into the frog's vocal sac, a deep skin fold starting in the front of the mouth which swells like a balloon during croaking and acts as an amplifier for the sound. In the Darwin frog, this sound bladder is greatly extended both laterally and in depth. It is probably the most original nursery in the animal kingdom, a place moreover that need not be laboriously constructed but is supplied by nature. Here the tadpoles live and draw nourishment from the plentiful reserves of their yolk-sacs and, in addition, take up oxygen and possibly even nutrients from their paternal host via the rich network of blood vessels in their large, delicately skinned tails. When they have completed their development and lost their tails, they leave their quarters by the same way they entered as eggs, through their father's mouth, as perfect little frogs.

Frogs demonstrate once again that when the brood is protected in early youth, the number of eggs produced is vastly reduced. The Darwin frog lays twenty or thirty eggs; the edible frog (*Rana esculenta*), in which concern for the propagation of the species is limited to the acts of spawning and fertilization, lays about ten thousand.

REPTILES

Zoologists combine in this class animals as different in appearance as lizards, snakes, tortoises, and crocodiles. Compared with the amphibia, which are still dependent on water or, at any rate, a humid environment, the reptiles have a more keratinized skin that protects them from desiccation and fits them for a life on land. However, in the millions of years since the Jurassic and Cretaceous periods, certain groups, notably the crocodiles and turtles, have once more returned to a watery habitat.

In regard to building or parental care, reptiles, whether aquatic or terrestrial, are even less remarkable than amphibia or fishes, though certain species must be given credit for guarding their eggs or for depositing them in natural cavities, or even in pits they dig themselves. However, the sea turtles do make a tremendous effort in the interest of their brood. Traversing vast oceans at egg-laying time, usually they swim to one particular beach which their parents and generations of their ancestors before them have visited for the same purpose and which they mysteriously manage to find again. There the females drag their heavy bodies up the sloping sand until they are above the tide line and laboriously dig pits with their paddle-shaped legs (pl. 69, p. 154). Having laid their eggs in the pits, they fill them in again, carefully smoothing the surface. When this is done the parent turtles return to the sea, leaving their offspring to fend for themselves. When the little turtles hatch, they dig themselves out without any help and quickly make for the sea. Alas, hordes of predators wait for them on the short way down and even more are stationed in the water. The worst enemy of the turtles, however, is man, who digs up the eggs from the sand and will, I fear, soon exterminate these primitive animals.

The brood-tending behavior of crocodiles is a noteworthy exception. Among them are species that protect their eggs by building primitive yet serviceable nests and actively tend their brood. Crocodiles' eggs are white, about the size of a hen or goose egg, and like these are protected by a hard calcareous shell. The Nile crocodiles choose flat sandy banks for egg-laying. A few meters from the water's edge they dig a pit, about fifty centimeters deep, with their front legs, pushing the sand to one side. The eggs are then laid in the pit and covered with the excavated sand; sometimes grass is added, which protects the eggs from excessive fluctuations of temperature. During the nearly three months it takes until the young crocodiles hatch, the female stays with the nest and, by growling fiercely, seeks to guard against covetous predators. The monitor lizard, in particular, is her enemy, for it is adept at digging up crocodile eggs and whisking them off to a quiet spot to eat at leisure. Just before they are ready to come out, the baby crocodiles make little croaking noises. Their mother then opens the pit and takes her family to the water. At first the young stay together, led by their mother. But soon they scatter and keep away from the older members of their species whose cannibalistic tendencies they have good reason to fear.

The salt-water crocodile (*Crocodylus porosus*), which lives on the coasts of southern Asia, and the American alligator (*Alligator mississipiensis*), which is found in all the southern states from Texas to Florida, build more impressive nests, sometimes a meter high, by carrying in their mouths branches, reeds, leaves, and decaying vegetable matter to a heap near the edge of a river or a swamp. Here again it is the female who stays near the nest. She not only guards and defends it, but, from time to time, using her tail, splashes it with water from the river or a nearby swamp water hole. The moisture accelerates fermentation in the decaying plant mass, and the heat engendered provides the uniformly high temperature in the interior of the heap which is needed for the development of the eggs.

Such nest-building activities of crocodiles are of particular interest because they lead without a break to the primitive nests of certain birds.

To some readers it may seem strange that I suggest with evident satisfaction a connection between crocodiles and birds. The clumsy crocodiles and the graceful birds on the wing appear to represent the greatest possible contrast. But as late in geological time as the Cretaceous period, there were flying reptiles (Pterosauria) of many kinds. Comparative anatomy and palaeontology prove without a doubt that phylogenetically birds evolved from ancestors resembling the reptiles. Hence, biologists are more gratified than surprised to find similarities of behavior between the two groups of vertebrates which today are so sharply divided.

BIRDS

Birds are hot-blooded creatures. Their normal body temperature, 41°–43° C. (105°–109.4° F.) would indicate extremely high fever in humans. Their high body temperature is an expression of a high metabolic rate, which is also reflected in their rapid movements and fast reactions. But basically what distinguishes birds—and mammals—from the cold-blooded vertebrates (fishes, amphibia, and reptiles), is not so much a high body temperature as the ability to keep the body temperature constant. It changes little even if the temperature of the air fluctuates violently, and remains the same in tropical heat and arctic frost. The body temperatures of lizards basking in hot sunshine may be every bit as high as those of "warm-blooded" birds and mammals. But when the outside temperature drops, the lizards—and other lower vertebrates—cool during prolonged periods of cold and fall into a state of torpor. They are not so much "cold-blooded" as *poikilothermous* (meaning "of variable temperature"). The various self-regulatory physiological mechanisms by which birds maintain and ensure an almost constant body temperature day and night, summer and winter, are not, or not yet fully, developed in the lower vertebrates. One indispensable aid to temperature maintenance in birds is their plumage. The material of feathers is keratine, like that of the scales covering snakes and lizards. Most birds still have scales on their legs, but on the rest of the body the former scales have changed their shape. They have become feathers, split up into very fine rays that fork and branch repeatedly into even finer subdivisions of microscopic size and interlock in such a manner that they form by ingenious devices a tight cover overlaying the fluffy down underneath and enclose a good many air pockets. Both keratine and air are poor conductors of heat and therefore excellent protection against heat loss. That is why we all love an eiderdown quilt on a cold winter's night.

Everybody knows that birds build nests and that they hatch their eggs with the warmth of their own bodies. It is less well known that some birds leave the hatching of their eggs to extraneous sources of heat. We have just seen that crocodiles do something of the kind when they heap up vegetable debris and keep it moist in order to utilize for the business of incubation the heat engendered by fermentation in a decaying plant mass.

Birds that build and regulate incubators

The same method, though much more highly developed, is encountered in the family of Megapodiidae (Greek for "large-footed"). They are gallinaceous birds and, as their name says, are characterized by unusually large feet. Feet of such size are vitally important from the moment of hatching, because the young birds are immediately faced with the difficult task of digging themselves out, unaided, from the depth of the breeding heap up to the light of day. And they will need these special feet again later in life to scratch their own breeding heaps together.

The megapodes live in Australia and New Guinea, and in the surrounding islands from the Nicobars to Polynesia. There are at least twenty different species, varying in size from that of a partridge to that of a turkey, all with inconspicuous plumage. Some live in dense bush, others in hot savanna country, and others in the mountains. Their nests and habits show most interesting adaptations to these different environmental conditions.

Plate 75. *Nest of mute swan. As female broods, male swan remains protectively close. (See p. 187.)*

Plate 76. *A male least bittern* (Ixobrychus minutus) *alights on his nest. (See p. 192.)*

Plate 77. An osprey flies up from his eyrie in the crown of a high pine tree. Easily recognizable are the boughs and branches he has utilized for his nest. Central Sweden. (See p. 197.)

Plate 78. Floating nest of a horned grebe (Podiceps auritus), *constructed from bits of aquatic plants and reeds, rises and falls with water level. (See p. 192.)*

Plate 79 a. Eyrie of a golden eagle built on rock face. Young bird on nest.

Plate 79 b. Golden eagles' eyrie, interior of nest with shattered egg. Both photographs made in Austrian Alps. (See p. 197.)

Plate 80. The nest of the great reed warbler (Acrocephalus arundi-naceus). *(See p. 197.)*

Plate 81. The reed warbler (Acro-cephalus scirpaceus) *on its nest. (See p. 197.)*

I shall start with the brush turkey (*Alectura lathami*), which lives in the forests of Australia's east coast. The cock chooses a site for his nest somewhere in deep shadow where, over a period of weeks, he scratches together an impressive mound of rain-soaked foliage, which in its upper layers also contains some soil. The expression "scratch together" may not give quite the correct picture. With his head turned away from the nesting site, he picks up his material with his foot and hurls it backward onto the growing heap. From time to time, the cock mounts the pile and stamps on it to make it compact. Eventually, the structure reaches a diameter of three to four meters and a height of about one and a half meters (pl. 71, p. 154, and pl. 82, p. 178). The hen that approaches at this time is not made welcome. The time is not yet ripe, not until the temperature inside the heap has settled down to a constant 35° C. (95° F.), or thereabouts, the warmth necessary for the development of the eggs. This temperature is usually exceeded in the beginning when fermentation proceeds rapidly. The amazing thing is that the cock checks the interior condition of the mound almost daily. Digging a hole deep enough for him to disappear into except for his tail, he repeatedly tests the temperature of the compost inside with his open beak. He takes some of it into his mouth and spits it out again when he withdraws his head. His behavior suggests that either his tongue or the inside of his beak contain highly sensitive temperature organs. If the pile is too hot, he leaves ventilation holes. If it is not hot enough, he adds further material suitable for fermentation and then closes the hole. When at last the compost is in the right condition, he calls the hen. She then lays her first egg into a deep hollow scratched into the heap and the cock closes it up with nesting material. This process is repeated once every two or three days over a period of several weeks. When she has laid the last of her eggs, the hen takes no further interest in the nest. But the cock remains fully occupied with the testing and regulating of his incubator until all of the eggs have completed their development. Each egg needs about nine to ten weeks from laying to hatching.

After leaving the egg, the chick must work its way to the surface of the mound, and this may take hours (pl. 70, p. 154). One might expect that the father, who has been toiling so hard to create and maintain the right conditions for the brood, would welcome the chick and take it under his wing. But this is not so. Even when he meets

one accidentally, he takes not the slightest notice. He does not seem to recognize it as his offspring. The chick itself seeks the nearest cover as fast as it can. It can flutter at once, and fly the next day—an extremely precocious nidifugous type. There is absolutely no contact between parents and offspring.

The egg of a brush turkey hen weighs three times as much as that of a domestic hen. The aborigines have long known that these mounds contain large and highly palatable eggs. Recently the plundering of the nests has become controlled in certain regions and licenses for their exploitation are issued by the government. It is fortunate that in these extensive forests many nests are never found.

The scrub fowl (*Megapodius freycinet*) is widespread in the Indo-Pacific regions. It is smaller than the brush turkey (roughly the size of a partridge), but its feet are exceptionally large, and it is the champion mound-builder. Its nests may reach a diameter of twelve meters and a height of five meters. No other bird constructs a nest of similar dimensions.

First prize for perseverance in strenuous work goes to

another species of megapodes, the mallee bird, or towan (*Leipoa ocellata*). These birds are also called "thermometer fowl" because they spend ten to eleven months of each year regulating the temperature of their nests. Their problem is that both daily and seasonal fluctuations of temperature are very great in the arid open bush of central Australia where they live. Moreover, foliage is scarce and any heap will soon be dried up by the sun and scattered by the wind. Prolonged and strenuous efforts are needed to produce a compost heap with a high and constant temperature.

Building begins in April or May, depending on the date of the first major autumn rains. (As the area lies south of the equator, autumn starts when spring comes to northern latitudes.) The mallee birds start by digging a large pit about a meter in depth for which they collect any twigs and leaves they find in the vicinity, at first close by, and later within an area of fifty meters. They fill the pit and heap up further vegetable material and a great deal of sand to form a mound on top of it, which is carefully smoothed. Both males and females take part in the work, the males doing most. Soon the compost below starts fermenting, but it takes four months until the desired constant temperature of 34° C. (93.2° F.) is achieved. This means that egg-laying can start around August. From then onward, the hens lay every four days or so. First, the cock digs a brood chamber in the compost (pl. 72, p. 155) and tests the temperature in the manner described for the brush turkey. Then the hen enters and checks the temperature herself. If she is not satisfied, the cock must find a more suitable place in the heap. When the egg is laid, the cock has to scratch the pit shut again.

After a preparatory period of four months, an incubation period of six to seven months follows until the hatching of the last chick. The adult birds, then, are occupied almost all year around with the business of building the incubator and tending it so the temperature of the interior, where the eggs of the clutch are lying close together, stays at an even 34° C. Temperature is checked almost daily and usually it is controlled to an accuracy of about 1°. The method for doing this changes with the season. In spring, it is sufficient to get rid of excess heat by making ventilation shafts and closing them at the right time. In summer, fermentation slows down but solar heat increases. To prevent overheating, the birds add to the sand layer of the mound. But when the heat of the sun gradu-

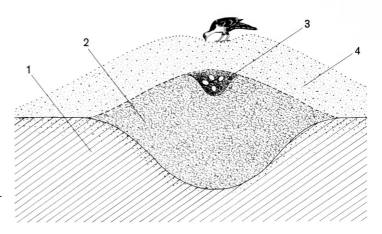

Fig. 74. Diagrammatic cross section of a brooding mound. (1) Natural soil; (2) the fermenting plant mass; (3) egg chamber; (4) sand.

ally penetrates deeper despite these precautions, they adopt more surprising and efficient countermeasures: they dismantle the dome in the cool hours of the morning, scratch a deep crater reaching close to the place where the eggs are, and spread out the sand. When it has cooled down, they throw it back into the hole and heap a thick layer of the old material on top for insulation (fig. 74). Each time this work takes two to three hours.

In autumn, when fermentation has ceased and solar heat declines, the dome is dismantled in the late hours of the morning and only a thin layer of sand is left on the eggs which are thus warmed by the midday sun. The sand that has been removed is spread in the sun, constantly turned over, and finally put back into the hole. This involves almost five hours of work, but it is effective. It is amazing how precisely the birds can adapt their activities to the situation and thereby succeed in holding the temperature in the egg chamber at an almost exact 34° C. during most of the time. Not until late autumn is there a slight fall in temperature.

We owe most of this information to the devoted research of the Australian ornithologist H. J. Frith, who studied the thermometer fowl in their natural habitats over many years. The birds soon learned to know him, and some pairs became so tame that he could watch them from close range. He and his co-workers were also able to introduce thermometers close to the eggs and check objectively the success of the birds' measures without disturbing them. The observers noted with surprise how well these methods worked every time and were anxious to probe further. Was the behavior of the parents really guided by the temperature they found during their checks of the interior? What would they do if the development of temperature was made to differ from its natural course?

The experimenters secretly introduced into the nest, in addition to thermometers, a heating instrument which could be controlled from outside. Obviously, such interference was completely outside the birds' experience, but most of the time their reactions were absolutely right. In spring, during the period of a slow fermentation-induced rise in temperature, the nest was usually opened once every two or three days. When the temperature was artificially increased, the birds opened the nest every day and soon were able to control the temperature once more. But when artificial heat was put on in summer, they could not perceive that the additional heat came from below and, in view of the season, took measures against overheating by the sun. They built the mound higher and higher to protect its brood chamber against what they took to be excessive solar heat; who knows how high it would have grown if a defect of the generator had not put an end to the experiment. In autumn, when normally the birds would have counteracted natural cooling by warming the sand and returning it to the nest, this operation was left undone when a check opening showed that the temperature was high enough. There is no doubt that the birds check the interior temperature and act accordingly. This probably applies to all megapodes. The various species test in the same manner with open beaks. In view of its purpose, the check is carried out only when the mound is opened, and only after a certain depth has been reached.

Frith tried another interesting experiment. He made a cut through a nest and closed the open section with a pane of glass so that the behavior of the chicks at hatching could be observed. As the mounds built by the thermometer fowls are more compact than those of brush turkeys, the chicks had to struggle for two to fifteen hours before they reached the surface. Exhausted, they sought cover in the bush. Only an hour later, however, they were able to run and flutter, and twenty-four hours later they could fly. But, as with all megapod fowl, there was no contact with the parents.

Though the cock does most of the digging, the performance of the hen is equally remarkable. She lays sixteen to thirty-three eggs in a season, and as a rule their combined weight is about three times her own body weight. As she lays an egg once every four days and each egg takes fifty days to incubate, the nest contains its full complement of about twelve eggs one month after laying starts. Thereafter, laying and hatching take place at the same rate until almost the end of the season. When no eggs are left, the parents separate for a short period of rest. After no more than one or two months, they meet again to start work afresh. Their union is for life.

Certain small birds on the volcanic islands of the region have found a considerably less strenuous way of caring for their offspring. On such islands one can watch thousands of birds leaving the forest at the breeding season. They collect at the foot of an active volcano and dig burrows in the loose tuff soil, which has a temperature of about 20° C. (68° F.). In an area of several square kilometers, nature offers these megapodes an enormous natural incubator while their less fortunate relatives elsewhere have to work hard to achieve the same end. Their behavior is, however, not unique. In other places, too, megapodes make use of volcanic heat in the vicinity of lava flows or hot springs.

Even simpler methods can be effective. Some species of megapodes living on islands leave the forest at breeding time and make their way to the coast where they lay their eggs singly into pits, which they have dug to a depth of a meter, and cover them with sand. Incubation is left to the sun.

On the beaches of coral islands, it may happen that the large turtles emerge from the sea at the same time as the megapod fowl emerge from the bush, and that both animals entrust their eggs to the sand side by side. Later in the year the newly hatched young go their different ways, the turtles into the sea, the birds into the bush. By then the parents of both have long since disappeared.

Both for the burying of eggs in sand—the most primitive form of parental care employed by megapodes—and for the construction of incubator mounds from leaves—the method used by other varieties—parallels can be found among the reptiles. It is a plausible theory that these primitive birds have carried over their methods of brood care from their cold-blooded phylogenetic ancestors, and have hung on to the old methods even though they have gradually developed the constant high body temperature that enables them to incubate eggs by the heat of their own bodies. But other scientists hold the reasoned view that, during a long transitional period, megapodes or their ancestors may have incubated their eggs themselves and that their present behavior is of secondary origin. If so, such a throwback may have been facilitated by the fact that the behavior of their ancestors was still "in their blood." No one can say with certainty what the course of development has been. But one thing is certain: under pressure from environmental conditions the thermometer fowl work harder for their nests than any other bird. I do not think we need to feel sorry for them. It may be assumed that they find satisfaction in following their instinctive drives.

Breeding without nest-building

Most people's general idea of a bird is of a feathered creature that builds a nest in which to hatch and raise its young. But this "general" idea is far too general. The breeding behavior of birds is considerably more varied.

The birds we have just been discussing build nests but do not sit on their eggs in them. Others sit on their eggs but do not build nests. In fact, quite a number of very different species, living in very different environments, belong to this group. I shall briefly discuss four examples to illustrate the variety of behavior and circumstances.

The guillemots (genus *Uria* of the Auk family) are typical marine birds; diving and paddling with their wings, they catch their food from the sea. At breeding time they congregate on one of the many famous bird islands, or on other cliffs along the coast of the northern hemisphere. Tens of thousands come to Greenland alone. Their favorite breeding sites are narrow rock ledges where eggs and young are well protected from predators. Usually the female lays a single egg on the naked rock. The danger of its rolling off the ledge is greatly reduced by its being pear-shaped. Even when it does move, it will roll in a narrow curve. There are no inviting nesting materials to be found on rocks or in the sea, and so it is understandable that these birds have chosen to do without.

It is less obvious why the fairy tern (*Gygis alba*) does not build a nest, because the tropical islands where this species has its home offer plenty of suitable nesting materials. Terns are related to the auks, though one might not guess it. The contrast to their clumsy cousins is great. With their long, narrow, and elegant wings, they are capable of sustained flight for longer periods than most other birds. From the other species of their family, most of which build nests, fairy terns differ by their color; they are the only pure white species among the terns. Seen streaking across a deep blue sea in the tropical sun, they appear almost ethereal.

When breeding time approaches, the bird migrates to land. Like the female auk, the female of the fairy tern lays a single egg, sometimes on bare rock, but usually on a branch often quite high above the ground, without even a vestige of a nest. Sometimes she chooses a fork, but more often a branch just wide enough to allow room for her feet on either side of her egg. There the egg must lie, and there the young bird has to remain until it can fly. The sharp claws and beak of even a very young bird help it to hang on with inborn skill and agility, as can be seen when a parent, taking flight, upsets its balance. A young fairy tern hardly ever falls off its perch. The dangers to eggs and baby birds are reduced by the unusual habits of the parents, both of whom brood: it has been observed that they change places only once every two or three days, and feed their offspring only twice a day. The parents bring as many as fifteen small fish or squids at a time, holding them carefully arranged across their beaks. In this way the risks inherent in taking off and alighting are considerably reduced. But it is not clear how these birds manage to go on fishing when they already carry several small marine creatures as prey in their beaks.

The ostrich (*Struthio camelus*), the largest living bird, inhabits a totally different environment. It is at home in the arid savannas and the deserts of Africa, even those almost devoid of vegetation. Ostriches do not build what could properly be called a nest, but the cock bird scratches a shallow depression in the ground in which he settles down. With his beak, he pushes under his body any eggs that a hen bird puts in front of his chest. As each hen may lay up to eight eggs and as polygamy is not infrequent, the brooding bird may be hard put to it to cover two or three times that number of large eggs with its body. This is quite a plethora of eggs compared with the output of auks and terns. Fortunately, perhaps, not all of them hatch. Both males and females take turns in sitting on the eggs. After six weeks, the young ostriches hatch and leave the breeding place at once, but the parents go on looking after them for quite some time.

From the point of view of nest-building, the hollow that the little ringed plover (*Charadrius dubius*) scratches into the ground is as primitive as the ostriches'. The eggs of this species are particularly well camouflaged by their color and markings (pl. 73, p. 156).

Simple nests

When a bird scratches a shallow depression into the ground, sits down in it, and turning this way and that, rounds it out and deepens it, this "cup-molding" process may be described as a primitive form of nest-building. The eggs are then placed in the hollow and brooded in it. As a further stage of development, the hard bottom of the depression is padded a little with blades of grass and other plant material. This is what most terns and gulls do and, provided it is done with care, the nest produced will be simple but very comfortable.

The nest of the mallard (*Anas platyrhynchos*), a duck common in North and Central America, in Europe, and in Asia, is of this type. Mallards usually choose nesting sites on the ground in a sheltered place among bushes, and they make a hollow which they line with loosely packed blades of grass or leaves, molded to fit the shape of the hollow. The female does not take a great deal of trouble collecting the nesting material, but picks up whatever she can reach from the nest with her outstretched neck. Once the eggs are laid, she improves the upholstery with down which she plucks from her own plumage. Brooding is exclusively the female's task. Whenever she has to leave the nest for a short while, she covers her eggs with leaves and down feathers, a precaution that not only preserves heat but also shields the eggs from the eyes of predators.

In the eider duck (*Somateria mollissima*), an inhabitant of northern climates, the down plumage is particularly soft and fluffy. The birds use their down feathers so lavishly for the padding of their nests (pl. 74 a and b, p. 156) that larger nesting colonies of these birds are regularly plundered without mercy by men who want the feathers for eiderdowns and pillows. In Norway and Iceland, large nesting sites are fenced off for the mutual advantage of birds and human beings. The "bird farmers" take the first two clutches with their nest feathers and leave the ducks their third clutch, three being the limit of the ducks' capacity of replacement. The birds appear to acquiesce in this treatment; at any rate, they come again year after year to nest in the fenced areas where they are safe from predators.

But let us return once more to the mallard duck. Occasionally she makes her nest not on the ground but in a tree. Because she cannot build a firm nest in the crown of a tree, she has to use the empty nest of a crow or a bird of prey as a basis for her simple home. Since the chicks leave the nest on the day they are hatched, when they cannot yet fly, they have to jump down. Their little bodies are so light that they usually come to no harm even when the drop is quite considerable. The mother duck then leads her whole flock—a female mallard lays eight to sixteen eggs—straight to the nearest water, where the chicks, which can swim though not fly, are safest. As sometimes the nest is quite a distance away from water, the young birds may have to perform a long cross-country march. After an incubation period of twenty-six days, all of the chicks normally hatch within a few hours. This is a great advantage as it enables them all to leave the nest with their mother after very little delay. The question of how this coordinated hatching is brought about turned out to have a simple answer. The chicks hear the first of their brothers and sisters pecking against the shell of their eggs, and this sound stimulates them to do likewise and free themselves.

The story I am about to tell shows that the choice of a breeding place is a matter of instinct rather than of reason or common sense. In the garden of the Zoological Institute of Munich University, situated near the main railway station in the heart of the city, we once built a few small islands about two meters long and almost as wide for the purpose of performing experiments with ants; each island was surrounded by a narrow ditch filled with water to prevent the ants from escaping. A pair of mallards from the river Isar happened to build a nest on an island that was unoccupied, and the mother brooded the eggs. The ducklings hatched on a Sunday when nobody from the Institute was about. The mother duck immediately started off with her family toward the river eighteen hundred meters away, right across the big city. Though she had arrived on the wing, she waddled in the right direction, an indication of her excellent sense of orientation. Fortunately, traffic was not too heavy because it was a Sunday. When she had covered about one-fifth of the distance, followed by her ducklings in single file, a policeman stopped the traffic on a central square, rang the Society for the Protection of Animals, and thus ensured a happy ending to this somewhat hazardous undertaking. Oskar Heinroth tells of another mallard duck that built her nest on the flat roof of a four-story building in Berlin. Her adventure ended sadly. The newly hatched ducklings, after leaving the nest, threw themselves off the edge of the roof and were killed by the fall. But this did not deter the duck from building her nest in the same spot in the following years.

The nest of the mute swan (*Cygnus olor*) is quite a simple affair. Yet the male builds the loosely arranged material into an impressive pile on which the female broods over the eggs, while her mate stands guard nearby (pl. 75, p. 173).

Perhaps it has occurred to the reader that the young of the birds chosen as our examples of species that build either no nests or only very primitive ones all leave the nest as soon as they have hatched. The connection is apt, for where the nest is of little importance to the brood, there is no point in taking a great deal of trouble. Little nidifugous chicks are born with downy feathers; they can walk and feed themselves, though they still need their parents to keep an eye on them, to guide them, and to warn them of danger. But the majority of birds are nidicolous. Their young are helpless little creatures (fig. 75) that must stay in the nest for days, or, in the case of large species, for weeks, before they are ready to leave it. In these circumstances, the building of a durable nest in which the young will find shelter and be protected against falling out is eminently worthwhile. The problem has been solved by different species of birds in a variety of ways, often amazingly skillful. It may be helpful to deal with a few general points before embarking on the description of individual examples.

Fig. 75. A nidicolous chick (left) and a nidifugous chick (right), both depicted a few hours after hatching.

General remarks on nests and nest-building in birds

Why is it that birds are often seen, but birds' nests only rarely? The nesting season is the most dangerous period not only for the brood but also for the parent birds who feel tied to it. It is a period in which they have most cause to fear the attacks of enemies. This is the reason why nest sites are well hidden in hedges and undergrowth, in tall grass, or in the leafy crowns of trees. One is often surprised at the many nests one discovers after the leaves have fallen in an area one knows well from frequent walks, nests one had never noticed before, although there must have been frequent comings and goings in the breeding season. Away from their nests, birds are much less reluctant to be seen because they can rely on their sharp senses and on their ability to fly. Small mammals are much less visible.

The choice of a nesting site is thus most important. Usually, but not always, it is the male that makes the choice. Each species prefers a particular nesting environment. The knowledge of what their breeding site should look like is theirs by instinct, handed down over untold generations.

The males of migratory species usually return from their winter quarters a few days before the females. At this time, a thrush that pours out its song from the same perch day after day is a familiar sight. When a bird has chosen a nesting area, its song tells all rivals that the site is taken and that they had better keep away. Just as fishes in many cases keep away rivals with threatening postures and a display of splendid colors, birds try to do the same, though more charmingly, by singing. In general, the claim is respected but, as with fish, quarrels and fights may occur. For unattached females, the bird's song has a different meaning. It tells them that an eligible male in possession of a territory is searching for a mate.

Courtship display, that is, the highly differentiated and ritualistic behavior of partners before mating, will not be dealt with here except where it is connected with special building activities. In some species, pairs unite for life. But the union also may last for one brood only, or for one year, or several years. Monogamy is the rule, but polygamy is not unusual.

Who actually builds the nest? Here, too, there is no single answer. In many species both partners build together. This is so for corvids, swallows, storks, and others.

Often there is some division of labor. Frequently, the male makes the rough basic structure while the female provides the soft interior lining. Or, as is the case with herons, the material is collected by the male and arranged by the female. Where the males do not tend the brood, as with the thrushes, pheasants, or ducks, it is usually the female alone who attends to the building. In other cases, the male bird is the sole builder. Among black woodpeckers (*Dryocopus martius*), only the male chisels a breeding cavity out of a tree trunk. Such a task, requiring major physical effort, does seem more fitting to the male partner. But wherever nest construction requires a particularly skillful job, the male bird works alone. Perhaps this is because very difficult work is handled better by one individual; why it is never the female bird that produces such works of art is a puzzling question.

The building tools of all birds are the same—their beaks and legs. It is different with nesting material. Birds that make holes to breed in do not have to collect nesting material, beyond a little padding for comfort. Those that build large nests on swaying branches must construct a strong framework. Hence, one cannot really make any generalizing remarks about nesting materials except to observe that hummingbirds do not carry heavy branches and eagles do not collect gossamer. The variety of materials used is almost infinite. Spiders' silk is an important raw material for some birds. Feathers, hair, straw, leaves, all sorts of fiber, twigs, branches, clay, mud, excrement, saliva, and much else are components of many nests. Each species of bird has its own rules, prescribed by heredity, for the choice of nesting material and the style and execution of its work.

The same is true for the way the material is brought to the site. A heron or stork carries each twig separately, while songbirds usually carry a whole beakful of plant material. Eagles carry the heavy branches they need for their eyries in their talons. A highly original method has been adopted by some species of the African genus of parrots *Agapornis* (from Greek *agape*, "love," and *ornis*, "bird), which are called "lovebirds" in English and "inseparables" in French, all names pointing to the closeness of their marital bond. They are frequently imported and bred in captivity. Many species of this genus carry their nesting material, consisting of little broken twigs and pieces of foliage, in their beaks, as most birds do, but some species put them into their plumage, especially between the feathers of the rump, thus keeping their beaks free. Clearly, they would lose much of their load during flight if their feathers were not specially adapted to the task and shaped somewhat like brushes, holding twigs and leaves by means of increased friction. An experiment made evident that not only the structure of their feathers but also their behavior is hereditary. Of two closely related species of lovebirds that were crossed by the breeder, one belonged to the group in which birds carry their nesting material in their beaks, the other to the one in which they stick them into their plumage. The offspring had inherited the urge to put twigs into their feathers from one of their parents, and the urge to carry them in their beaks from the other. These two hereditary urges were so compulsive that the hybrids were unable to let go of the twigs they had put into their feathers. This conflict interfered so strongly with the whole process of collecting material that nest-building failed.

If searching and gathering nesting material is found burdensome, it is often possible to steal it. H. O. Wagner, who studied the nest-building behavior of the white-eared hummingbird (*Hylocharis leucotis*) in Mexico, frequently noticed such thieving actions. While the female sat on the nest and incubated her two young, which had hatched only a few days before, a violet-eared hummingbird (genus *Colibri*) came whirring along and pulled out some building material. In the course of three days, it had made a sizeable hole in the thick nest wall. After another four days, only a tattered remnant of a nest was left, harboring one baby bird, while the other lay dead on the ground. On the following day, the second baby had disappeared along with the remains of the nest. The strange thing was that the mother did not seem to react by defending her nest. Though she saw what was going on, she did not seem to grasp its significance. In another case, in which the nest was still under construction, the thief carried away the nesting material as fast as the white-eared hummingbird brought it along, until the first bird gave up her frustrating efforts. Similar behavior has been observed with penduline titmice and other species.

It is, of course, even simpler to steal a whole nest. Admittedly, a nest cannot be carried away, but it can be usurped by various methods. N. E. Collias, in Ohio, observed how a starling caught a flicker by the tail and chased it from its newly constructed nest cavity; later, the starling used the hole for its own brood. The family of Tyrannidae (tyrant flycatchers), widespread in North and South America, derives its name from the aggressiveness of many of its members which seek to repulse trespassers from their territory by fierce attacks and loud scolding cries, even when these intruders are much bigger than they are themselves. One of these species, the piratic flycatcher (*Legatus leucophaius*), uses this habit nefariously. When they find a nest that appeals to them, one of the tyrant couple starts a quarrel with the birds occupying the nest. While the owners of the desired accommodation fly off in hot pursuit of the piratic intruder, the other partner slips in, throws out the original owners' eggs, and usurps the nest. The European nuthatch (*Sitta europaea*) is less brutal but no less efficient when it comes to stealing the breeding nest of a starling. It uses the same method by which it generally protects its home from predators larger than itself. In the absence of the starling, this interloper narrows the nest entrance by plastering on clay and soil so that only a very small bird like itself can enter.

Just as there is no general rule about the division of labor in building, there is no uniformly accepted behavior for brooding. In many species, father and mother relieve each other (as do pigeons, cranes, storks, starlings, and orioles). In other species, only the female sits on the eggs (ducks, owls, crows, titmice, finches, hummingbirds); in still others, only the male (rhea, dotterel, northern phalarope). At breeding time, it is easy to see whose job it is: the brooding partner has a bare patch on its underside, without feathers, which is red because it is rich in blood vessels. If only one of the partners broods, only this partner has a brood patch. If both do, then both have them. Brood patches have a high temperature: they make primitive but effective heating devices for the eggs.

In nidifugous birds, the nest is of no importance after the young have hatched. In nidicolous birds, it becomes the sheltering home for the helpless babies. Its function ceases only when the little birds are fledged, and therefore it has to be constructed differently. It must be less easy to climb out of, and it must provide the young with better protection against heat loss than a shallow, dish-shaped nest.

Cup-nests

To achieve better shelter for their young, the birds must build a nest in the shape of a deep bowl or cup in place of a nest fashioned like a flat cushion or a shallow dish. It means that the side walls have to be raised. Simple nests of this kind are built, for instance, by the bitterns. These birds, members of the heron family, live near lakes or slow-moving waters with reeds growing near the shore and feed mainly on fish, frogs, and insects. Four species of bittern of the genus *Botaurus* occur in North America, Eurasia, and Australia. The least bittern (*Ixobrychus minutus*), which is about half their size, inhabits Europe, Africa, and Australia. A nest of this species is shown in plate 76 (p. 173). The photograph depicts a male arriving with measured, quiet steps to relieve his mate, one leg still on a reed, the other already on the nest. The bitterns are well camouflaged by their coloration, and this protection is made more effective by their mode of behavior and movement. Among least bitterns, it is the male that starts nest-building, but later the work is carried out jointly by both partners. These birds also relieve each other in the business of brooding. In *Botaurus*, only the female is engaged in nest-building and incubation, a fact that illustrates how greatly behavior may differ between species that are very closely related.

The nest of the least bittern is usually sited just above the surface of the water and bent or broken reeds frequently serve as a base. It is built from reeds laid one upon the other or interlaced in such a manner that they give each other mutual support (pl. 76). The interior is padded with soft material, such as the tender tips of bulrushes.

The nesting habit of the horned grebe (*Podiceps auritus*) is even more aquatic. The nest, which is built from aquatic plants, bits of reed, and the like, floats on the water, rising and falling with its surface level (pl. 78, p. 174).

Herons' nests are larger; they are usually built on trees, but their construction is similar. Most species breed in colonies. As their building material includes branches of considerable size, the nests of a heronry are quite durable and can be used again another year. When the gray herons (*Ardea cinerea*) of the Old World return from the south to their familiar breeding grounds, the first males to arrive take possession of the largest nests. Late-comers must be satisfied with smaller accommodations. In bird colonies, no less than in human settlements, close vicinity stimulates comparison with neighbors and arouses covetousness.

Plate 83. A lesser whitethroat (Sylvia curruca) *brooding on her eggs. (See p. 197.)*

Plate 84. *A leaf warbler* (Phylloscopus collybita) *in front of lateral entrance to its spherical nest, built near the ground. In its natural leafy surroundings, it is hardly recognizable as a nest.* (See p. 202.)

Plate 85. *Wrens do not only nest in undergrowth, but also on overhanging banks of streams or ditches. It is feeding time.* (See p. 202.)

Plate 86 (facing right). *In central and northern Europe, storks build their eyries on the roofs of houses and on church steeples. Originally they nested in trees, as shown here, and as still found in most of southern Europe. This eyrie photographed in Macedonia.* (See p. 197.)

194

Plate 87. *A penduline titmouse with food at the entrance tube of its nest. (See p. 206.)*

The larger the bird, the sturdier the nest, but the building technique may remain essentially unchanged. Ospreys (*Pandion haliaëtus,* pl. 77, p. 174) and golden eagles (*Aquila chrysaëtos,* pl. 79 a, p. 175) collect branches and sticks of one to two meters in length, as well as smaller twigs, for the construction of their imposing eyries which they erect in the crowns of trees, on rocks, or on telephone poles. The insides of their nests are more refined and are lined with brushwood, moss, or the like (pl. 79 b, p. 175).

An eyrie is used again year after year, enlarged and repaired with new branches. One particular eyrie of a bald eagle (*Haliaëtus leucocephalus,* the national symbol of the United States), was used for thirty-six years, until the tree on which it rested was felled by a storm. In the same way, the nests of the European white stork (pl. 86, p. 195) increase in size from year to year, and are used over and over again, though not necessarily by the same pair. One storks' nest, in a tree in Hungary, grew to a diameter of two meters and a height of two and a half meters over the course of many years.

Such impressive birds are not an everyday sight. If we look at the more familiar world of small songbirds, we naturally find more delicate building materials and also a more careful workmanship. The reed warbler (*Acrocephalus scirpaceus*) usually nests in the bulrushes and anchors its nest, which is plaited from grasses, strips torn from reeds, and other dry thin building materials, to adjacent reeds (pl. 81, p. 176). The hollow of the deep basket is lined with the panicles of rushes and similar soft padding material. The nests of the great reed warbler (*Acrocephalus arundinaceus,* see pl. 80, p. 176) and those of the closely related warblers of the genus *Sylvia* (family Sylviidae) are fashioned similarly (pl. 83, p. 193). Observation has shown that their construction follows certain basic patterns.

When the first pieces of grass and stalks have been collected and have found reasonable support in a branch fork or in the dense entanglement of a climbing vine, the bird sits down and makes rotating movements with its body. These are the same kind of "cup-molding" movements which the male ostrich makes in the sand of the desert—with the difference that this is all the nest-building he does. *Sylvia* warblers sometimes make these molding movements even before they have collected a single straw, thereby taking possession, symbolically, of the site of their choice. As soon as more nesting material is accumulated, these molding movements achieve a visible result and really create a hollow. Lying in this hollow, the bird then kicks out with its legs, pushing them backward in quick succession against the raised side wall to compact it. It achieves the same effect by pressing its chest against a part of the wall. Repeated turning movements produce a regular circular shape and equal compaction all around. But to impart better built-in strength to the growing little basket, the bird must resort to a plaiting technique, which need not be complicated. Crouching in the nest hollow, with its breast slightly retracted, the small builder plucks at a straw or a stalk and pushes it in again at a distance of a few centimeters. Or the bird pushes a thin twig through the nest wall with tremulous movements of its beak, twisting it back and seeking to anchor it. To make the rim of the nest strong and smooth, the bird often collects the fine threads of spiders' silk. When it returns with spiders' webs, it first brushes them off its beak onto twigs where they attach themselves. They can then be drawn out to fine threads that are shuttled back and forth across the rim, firming the structure.

The inside is lined with soft material such as hairs, feathers, spiders' or insects' webs, or whatever else is handy. In this group of songbirds, such work is usually carried out jointly. However, the blackcap (*Sylvia atricapilla*), a bird native to most parts of Europe and beyond and well loved because of its beautiful song, behaves somewhat differently. The males return from their winter quarters ahead of the females and start building several nests in their territories, but leave them half-finished. When a female arrives, she chooses among the various skeletal structures that the male offers to her. She helps him to finish the one she likes best, and now works with a zeal exceeding his own.

To watch a hummingbird in the field with its flashing metallic colors and its marvelous agility in flying is pure delight. Suddenly a bird appears as if by magic. It hovers seemingly motionless in the air before a flower, its only movement the whirring of the wings, imperceptible because of its speed. With its long beak, it collects nectar from the depth of a flower's corolla. Setting off backward, it rushes to the next goal in an elegant arc. Its main food, apart from sweet nectar, are small spiders and insects picked from the flowers or caught in flight. Hummingbirds are the tiniest creatures in the world of birds. The smallest among them is no bigger than a large bumble-bee.

Such a small size presents a special problem to a warm-blooded animal. The smaller the body, the larger is its surface in relation to its volume. And the larger the relative size of the surface, the more heat is lost to the atmosphere. This is a well-known fact of everyday life. A large dumpling will take longer to cool on our plate than a small one, and if we are in a hurry, we will cut the large dumpling into small pieces to speed up the cooling process. Very small warm-blooded animals in a cold climate have to eat a great deal at short intervals to provide the fuel that maintains the heat of their bodies. In our context, this means that hummingbirds must build nests which can keep their extremely small young sufficiently warm. Hummingbirds (family Trochilidae) are found only in the New World. They occur in both North and South America, but are by no means restricted to the tropics. They may be encountered in high northern latitudes and can also be found in mountains.

The females are usually inconspicuous, but the males frequently possess a beautiful, multicolored, iridescent plumage and may have long tail feathers. When they court a female, they display their colors in aerobatic flights. But once the pairing is over, they are not very considerate fathers. They make themselves scarce and leave it to the female to build a nest and raise the brood.

Hummingbirds build open, cup-shaped nests. These are made from particularly fine material and built with great care to a high standard of density. In cooler climates, their bottoms are made very thick and the side walls, too, are of considerable dimensions. It is difficult to watch a hummingbird building its nest, but H.O. Wagner was able to study the activities of the white-eared hummingbird (*Hylocharis leucotis*) during a ten-year stay in Mexico, and gave the following description. A female, which had chosen as her nesting site the fork of two thin branches not very high above the ground, started by bringing spiders' webs in her beak, fastening them on the fork in stationary whirring flight, and stretching them into taut threads, thus building a foundation from spiders' silk. She then collected the fine plant hairs of oak gall (caused by sawflies), which were common in the neighborhood, and sheeps' wool caught on thorny shrubs. Whatever the bird brought along in her beak, she pushed into the growing nest wall in whirring flight. Occasionally she brought other light materials, such as small dry leaves or lichen, but most often spiders' webs. These were used constantly to join various parts together and to impart firmness and flexibility to the whole structure. The bottom alone, made mostly of moss, took about a week of unremitting work, and the raising of the walls another week or two. Simultaneously the cup was molded by the bird, sitting in it, turning this way and that, and constantly pushing the material with her beak into the right position. Finally, a thick ring of spiders' webs was built up in many layers on the upper rim of the nest. More spiders' webs connected the nest with the nearby leaves and twigs, though possibly they were carried there by the wind (fig. 76). The hollow was not lined. It was soft enough as it was.

A hummingbird clutch consists rarely of more than two eggs. When the mother sits in the deep cup, only her tail and head, with the beak pointing upward, is visible. Between the thick soft bottom of the nest and the down of this live little heater, it is cozy enough. After two to three weeks, the young hatch. They stay in the nest, where they are well looked after, until about three weeks later. Then they are ready to reach the next branches in flight. The mother feeds them for another few days, but soon they are quite independent.

During cold spells, the walls of the nest, thick and dense though they are, can slow down but not prevent a heavy loss of heat during the unavoidable nocturnal feeding interval. The hummingbirds cope with this by a remarkable physiological peculiarity. Instead of trying in vain to keep their temperature constant and thereby burning up all their fuel, they allow their body temperature to drop and so, by reducing temporarily their metabolic rate, they fall into a state of torpor. In this respect they behave like cold-blooded animals and thus manage to withstand periods of really severe weather. As soon as the sun reappears, they warm up their bodies with shivering movements, and soon they whir away once more in search of food.

More than three hundred different species of hummingbirds are known. All over the world the larger zoological museums pride themselves on their collections not only of the birds themselves, but of the great variety of their nests as well. Though their material is fairly uniform, the sites may vary considerably. An open cup-nest

Fig. 76. A white-eared humming-bird on her nest, situated in a small hollow of an embankment. Spiders' webs are an important nesting material. Cobweb threads are here shown stretching between the nest and the foliage and small twigs around it, as if wafted there by wind.

is liable to collect rain when the sitting bird leaves it even for a short time. Many hummingbird species try to guard against this danger by choosing a covered site. They build their nests under large leaves, or below an overhanging rock, or in a cave. Such sites have the added advantages of reducing loss of heat and protecting against excessive insolation.

A roof over one's head

In Europe, the wren (*Troglodytes troglodytes*) is one of the smallest members of the bird community. (It is represented in North America by the winter wren.) It lives in different environments but prefers dense undergrowth, scrub, piles of slash, or similar hiding places. There it flits about so fast in short flights, hops, or runs, that it is often taken for a mouse. If it can be seen at all clearly, it can easily be recognized by its posture and its short upright tail (fig. 77). Males and females have the same plumage. Only in northern regions do they tend to migrate south in winter. In temperate climates they are permanent residents, and though their food consists chiefly of insects and spiders, they try to get through the winter by foraging for scant food in nooks and crannies. Their cheerful songs and their rattling warning calls can be heard even in the snow, but more so, naturally, when nesting time approaches. Nest-building is started by the male alone.

A wren's nest is spherical and has a small lateral opening (fig. 77). The fact that the bird thus builds a roof over its head represents an evolutionary advance in comparison with open nests. The site may vary considerably, but it is frequently hidden under exposed roots or the like. As the plant material for the construction of the nest is taken from the immediate vicinity, it harmonizes with its surroundings and provides a good camouflage. Frequently, the wren chooses a site under some overhang (pl. 85, p. 194). Leaf warblers (genus *Phylloscopus*) build similar spherical nests with lateral openings in dense undergrowth (pl. 84, p. 194).

Like the blackcap, the male wren builds several rough nests. These he shows to an interested female who chooses the one she prefers and proceeds to finish it. But the wren is not a faithful spouse. While his new bride is busy decorating the interior of the chosen nest with feathers and other soft materials, he invites other females to look at his remaining nests. He is definitely a polygamist. As a result of this habit, many males fail to find a mate. Possibly they are the ones whose nests were not good enough, so that this kind of selection may serve to maintain, or even improve, building standards. The male wren cares little for his brood, though occasionally he helps with the feeding.

Wrens often use the breeding nest after the young have flown, as well as some of their half-finished nests, as a place to roost, especially in cold weather. Most people will not think this remarkable. It is often believed that birds build their nests not for breeding alone, but also as sleeping places. However, such behavior is rare. During incubation there is usually no room for the second bird. In any case, most species look for roosting places in the branches of bushes or trees, or in other sheltered spots. Some sleep on the ground, and aquatic birds prefer

Fig. 77. The spherical nest of the wren can usually be found close to the ground, in dense undergrowth, where it is difficult to discover.

reeds or the water itself where they feel safest. Nevertheless, some birds, after breeding is over, use their nests as sleeping places or even build nests especially for this purpose. These exceptions include the house sparrow (*Passer domesticus,* usually called the "English" sparrow in America) and the European tree sparrow (*Passer montanus*) as well as the magpie, some woodpeckers, and, as just mentioned, the wren.

The real home of the wrens is in tropical America. The family of Troglotydidae is represented in the two halves of the continent by over sixty species, of which a single one has penetrated into Europe and Asia. They all have many common traits; the habit of building nests for sleeping is widespread in America too, and, in some species, it is of considerable importance. This is so with the cactus wren of the United States (*Campylorhynchus brunneicapillus*), which builds its nest as a proper home for the whole family where they can find shelter from rain and cold all the year around and where they can all sleep at night. Once the young birds have grown up, they build their own nests for protection in winter. Some wren species have not adopted the "modern" roofed style of building and continue to build open cup-nests.

A hanging nest protects against unwelcome guests. The wrens are mainly American birds and have only one representative in Europe. Penduline titmice, on the other hand, have only one species in the New World, which lives in the southern parts of Pacific North America, while ten species are known from Europe, Asia, and Africa. Penduline titmice (Remizidae) are closely related to other titmice and formerly were ascribed to the same family (Paridae).

The nests of penduline titmice are closed at the top and have a lateral entrance hole, like those of the wrens. But their choice of nesting site is more advanced. They attach their nests to the ends of thin twigs, often the shoots of willow trees, in such a way that they swing freely in space. This renders their homes virtually inaccessible to predators hunting in the crowns of trees.

These bag-shaped nests have long been greatly admired because of their exquisite workmanship. They are so strong that in eastern Europe, where these birds are common, children sometimes wear nests instead of slippers; the Masai tribe use the nests of an East African species as purses. How does the bird go about fashioning this handsome and durable structure?

One such nest, made by *Remiz pendulinus,* is shown in plate 88. This penduline titmouse, which occurs in Germany but is rather rare, likes water and usually nests in stands of willow, poplar, or birch. Sometimes it suspends its nest directly over a water surface.

At the onset of spring the male starts building, alone. He uses long tough strands of fibrous material, such as grass, bast, adventitious roots of willows, and occasionally hair, to make the basic structure. Fluttering around the tip of a hanging branch, he attaches to it a strand, which he carries in his beak, and secures this by winding it round the twig. Additional threads are fastened in the same manner or interlaced and interwoven with material already there. The bird uses its beak to push the thread

ends through the meshes. This produces at first a kind of downward-pointing strip which is later divided into two parts that are lengthened, broadened, and joined together again further down. The resulting shape is like a little basket with a handle (fig. 78, page 206). After that, a wall is added at the back, and the wide opening in front is closed, except for a small entrance hole.

However, this description is not yet complete. The nest is more than just a basket plaited from fibrous strands. Its construction is reminiscent of that of an oriental carpet. In the same way as a carpet weaver knots short

Plate 88. The nest of a penduline titmouse. Specimen in the Naturhistorisches Museum, Braunschweig.

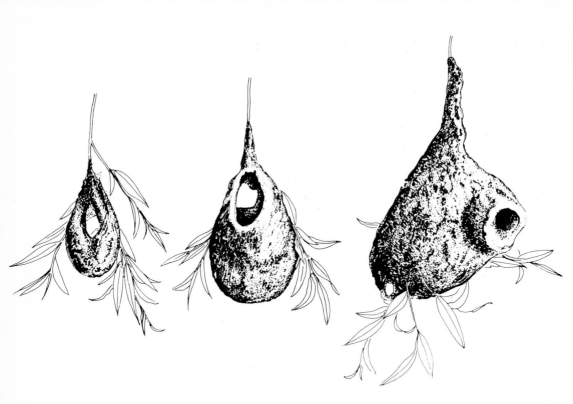

Fig. 78. Two stages in building
and a finished nest of a penduline
titmouse.

strands of wool into a coarser basic fabric, so the penduline titmouse uses short vegetable fibers, especially the fluffy seed hairs of willows and poplars, in addition to the long ones, and works them into the meshes and gaps of the original fabric to make it thick, felted, and durable (plate 87, page 196). It is hardly surprising that such a nest takes three to four weeks to build. As soon as it nears completion, the male tries to win a female. When he has secured a bride, she helps him to finish the nest and lines it with soft plant down. The female alone shoulders the task of brooding, leaving the male free to build a second nest and woo another female.

As one would expect, different species have their peculiarities, including special methods for making the brooding space even more secure. The method used by the verdin (*Auriparus flaviceps*), the only American Remizidae species, is rough but effective. It builds its round nest from thorny twigs, which are not difficult to find in the semi-deserts where it lives. The method of the African penduline tit (*Anthoscopus caroli*) is more sophisticated. When building its nest, it attaches a flap, extending toward the back, to the bottom part of the

206

entrance hole. On leaving the nest, the bird pulls this flap up and thereby closes the entrance door. The reader will probably remember another door that can be shut, in the discussion of trap-door spiders.

Weaverbirds. The building skills of the penduline titmice are less visually striking and, therefore, less known than the more conspicuous achievements of the weaverbirds. One reason for their greater fame, no doubt, is the fact that they are very numerous and occasionally breed in large colonies. Their many nests hang from the trees like large fruit (pl. 89, p. 213). Other species build communal nests that are larger than any other bird nests found in the crowns of trees (pl. 93 a and b, p. 216). We will consider these structures in a moment, but first some observations on the kind of birds we are discussing. One member of the weaverbird family (Ploceidae) is known to everybody—I refer to the ubiquitous house sparrow ("English" sparrow to Americans) that thrives near human habitations and has followed man wherever he has gone.* Its untidy nest has certainly not contributed to the family's fame as weaver. However, the sparrow group (Passerinae) forms only one of ten subfamilies of greatly varying skill, the most accomplished of which is that of the weaverbirds proper (Ploceinae). Most of its seventy species, which are roughly the size of sparrows or starlings, live in Africa, others in southern Asia.

Their technique of nest-building does not suggest knotted carpets, as with the penduline titmice; but some are remarkably clever weavers and produce structures of great density and durability whose shapes are often similar to those of the penduline titmice. The nests are always closed at the top. The opening is either on the side or at the bottom. In the latter case, the opening is often placed at the end of a long flight tube. Such tubes are frequently encountered in nests suspended from the ends of thin branches. While the roof provides shelter against rain and tropical sun, the flight tube is designed as protection against dangerous tree snakes.

In plate 91 a and b (p. 214), the nest of Cassin's weaverbird (*Malimbus cassini*) is illustrated. The working method of this bird, which is one of the best weavers, has been carefully studied. The technique is the same in

* The "English" sparrow was not native to North America, but was intentionally introduced in the nineteenth century.

many species, but not all of them work with the same care and precision.

Here, too, nest-building is chiefly the job of the male. At the start, the female is not in evidence, though later she will see to the interior decoration. The choice of building material is dictated by availability, but it must consist of thin strips that are both flexible and strong, such as blades of grass or parts of palm fronds. The bird grasps the edge of a leaf or blade with its beak, and in flying away rips off a thin strip of some length. This strip and others like it are twisted around the end of a twig and woven together with further thin pieces (fig. 79, top left). As with the penduline titmouse, this initial hanging

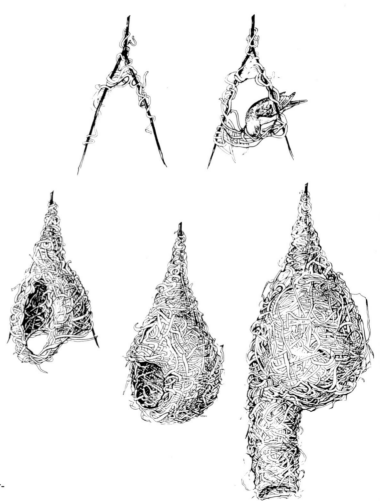

Fig. 79. How the nest of a weaver-bird takes shape.

208

section is divided into two parts that are joined together again lower down to form a ring (fig. 79, top right). This ring is widened laterally and extended to a brood chamber on one side and an antechamber on the other, which contains the opening that leads into a downward-pointing flight tube (fig. 79, bottom). The bird works more or less like a basket weaver, and partly even like a weaver with a loom. But the strands the bird has to work with are shorter than the warp of a weaver's loom, and it has to secure the ends repeatedly by pushing them into the meshes of the existing fabric or by tying them into it. It does this with its beak; often it uses its foot to hold the fabric steady (fig. 80). The method of fastening

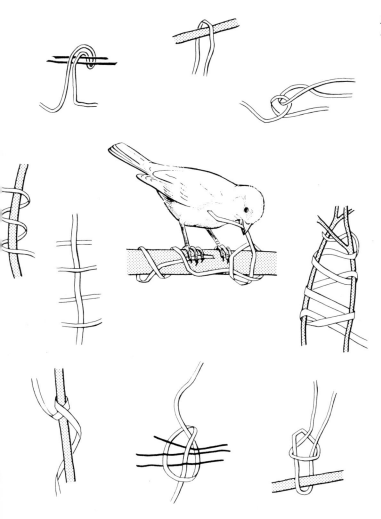

Fig. 80. A few examples of the knots and loops of a weaverbird.

varies with the circumstances. In the simplest case, the strand is placed in a loop around a twig or around another strand. Sometimes it is pulled through an existing loop, or even tied into a knot, either by making several spiral turns or by intricate interlacing. Some examples are given in figure 80. The bird is adept in pushing a strand through the network and pulling it out at the next hole. Once a solid basis for the final shape has been achieved, the density and firmness of the wall can be strengthened by proper weaving according to the principle of warp and woof, as can be clearly seen in plate 91 b (p. 214). Here the various threads run neatly at right angles to each other and diagonally to the direction of the nest tube, which is thus made both flexible and strong.

Though the male weaverbird is anxious to make his fabric strong, he avoids pulling his knots very tight because, like Penelope of old, he may want to undo what he has laboriously constructed. The reason for this is curious. As soon as the nest has taken shape, the weaverbird tries to find a mate. The females see the nest and understand the meaning of the male's fluttering display around the nest entrance, but they are very fastidious and make considerable demands on the quality of the home that is offered them. If the male has not been able to find a female that will accept his nest within a week or so, he himself demolishes the fabric that he has taken such pains to make. In that case it is obviously an advantage if the knots are not tied too firmly. He then tries to produce a better nest in the same spot. If this is approved of by a female, she takes over decorating its interior with grass and other soft materials.

A male wren will always build a number of nests for the female to choose from. A weaverbird makes a single dwelling, and if the females do not like it he demolishes it and tries anew. With both birds, the work produced may vary greatly in quality. Differences become even more apparent if one compares the nests of old and young weaverbirds. Young birds start building in the first year of their lives before they are sexually mature. The nests they make are rather crude affairs, presumably because their building instincts have not yet matured. Many instincts develop fully only when they are needed. However, these early playful attempts at building appear to teach the young birds how to handle their material and help them to develop their skill. To a certain extent the building activities of these birds result in individual creations. There is a difference in kind between them and the activities of fishes and amphibia, or those of insects and spiders whose highly developed instincts cause their first nest, or their first web, to be a work of perfection.

Weaverbirds tend to breed in colonies. Clustered on large isolated trees in the African savannas, their nests often catch the eye because of their numbers alone, even where they do not show the long conspicuous flight tubes of Cassin's weaver (pl. 91 a, p. 214). In plate 92 (p. 215), a colony of oryx weaverbirds is shown in detail. Their retort-shaped nests with the short flight tubes pointing downward are of a different design. Plate 90 a (p. 213) illustrates a male textor weaverbird (*Ploceus cucculatus*) engaged in weaving, and in plate 90 b we see him in front of his half-finished nest, trying to attract a female by a courtship display of whirring flight and song. In this and many other species, the male plumage is colorful while that of the female is drab and resembles that of a female house sparrow.

Communal nests

In other species of weaverbirds, the urge for a social life is so strong that it leads to the building of communal nests for several families. The buffalo weaver (*Bubalornis albirostris*), which lives in the African thornbush, is one of the largest weaverbirds—about the size of a starling. Its use of thorny shoots and twigs for its nesting structure demonstrates once again how similar environmental opportunities may lead to analogous behavior. In America, the verdin barricades its nest with thorns as with a barbed-wire entanglement, and in Africa, the buffalo weaver does exactly the same thing. A group of paired birds starts building their spiked nests close to each other in the branches of a tree as if they merely intended to form a loose colony of neighbors. Soon the gaps between the nests are bridged with more thorny twigs, and this eventually results in the formation of one large unified structure that may reach a diameter of two to three meters. It presents a forbidding front to any outsiders contemplating interference with the communal home. There are separate brooding areas inside, each with an individual entrance from below. Within the thorny pile, the nests are carefully padded by the females with bits of grass to form cozy little homes that are also used for sleeping.

The communal nest of the sociable weaverbird (*Philetairus socius*), a bird closely related to our sparrows, is even bigger. Several pairs start from the outset to build a joint nest attached to a stout branch. Like the wasps, they build their house from the top downward, using twigs and the strong grasses of the dry South African savannas. They, too, establish individual nests within the communal structure that usually houses some twenty to thirty couples, living, as far as can be ascertained, in exemplary monogamy. The community home is extended year after year until some nests reach diameters of about five meters (fig. 81 and pl. 93 a and b, p. 216). The flight holes are all on the underside (pl. 93 b). As many as 125 have been counted on a single nest. A communal nest of weaverbirds is probably the most conspicuous birds' nest in existence. There is no need for concealment to protect it from the eyes of enemies, a necessity which affects the choice of nesting site in so many other species. The birds do, however, face *one* danger, a hazard that increases as their fortress home grows. The time may come when the supporting branch is no longer able to carry its weight, and the whole communal structure, built up so laboriously over many years, suddenly falls to the ground and collapses. Fortunately, such accidents are rare and usually occur when a home has been occupied

Fig. 81. The colony of the sociable weaver (Philetairus socius) *may reach a diameter of about 5 m. It contains under one roof twenty to thirty pairs in separate nests. Above, drawing of a sociable weaver.*

Plate 89. Nests of weaverbirds hang from tree like giant fruits. A colony of Pseudonigrita arnaudi in Tanganyika. (See p. 207.)

Plate 90 a. A male textor weaverbird, engaged in weaving.

Plate 90 b. Weaverbird at the entrance of his half-finished nest attempts to attract a female. Photographed in southwest Africa. (See p. 211.)

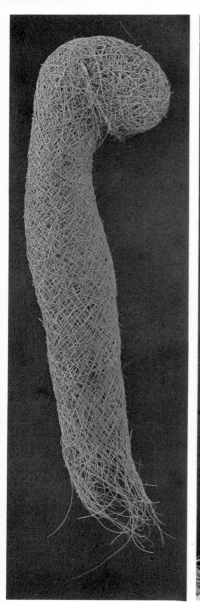

Plate 91 a. Nest of a weaverbird (Malimbus cassini) *removed from the branch to which its upper end had been attached.*

Plate 91 b. Detail in natural size of a weaverbird's nest. (See pp. 207, 210, 211.)

Plate 92 (facing right). Detail from a colony of oryx weaverbirds (Euplectes orix), *mainly found in South Africa. (See p. 211.)*

Plate 93 a. A giant nest of sociable weaverbirds seen from below, showing the many flight holes.

Plate 93 b. Colony of sociable weaverbirds. Sideview. (See pp. 207, 212.)

216

for many years, having given protection and shelter, like a system of deep caves, to generations of weaver-birds.

Birds as tenants

It is not unusual for other birds to establish their homes in the communal nests of the sociable weavers. Small parrots, pygmy falcons, and other species will occupy empty nesting chambers. They might be called subtenants, or squatters.

Such behavior is by no means isolated. Starlings, sparrows, and other small birds occasionally build their nests in the tangle of branches of an eagle's eyrie, even while the eagle is in residence, and the piled-up masses of twigs in storks' nests are also used at times as nesting sites by small birds. The owners of the large nests do not seem to mind, and their small guests are well protected from predators.

Cave-breeding birds in the tropics find excellent sheltered breeding sites in the fortified homes of the highly aggressive termites. Fifty different species of birds have been noted to hack through the outer armorlike plating of the nests of soil or tree termites, creating holes just large enough to slip through and build nests inside. Ants' nests are also popular breeding sites. Woodpeckers, for example, like to construct their breeding cavity in the carton nests of *Crematogaster* and related species. The birds not only live rent-free, but they act as usurpers and frequently destroy large numbers of the rightful inhabitants. Strangely enough, these ants never attack the woodpeckers and their brood.

Some weaverbirds and other tropical species seek the protection of wasps and hornets. Though they do not try actually to live in the nests of these aggressive stinging insects, they build their own nests in the immediate vicinity. Figure 82 depicts the nest of a warbler from the South Pacific (*Gerygone*) hanging on the tip of a branch next to that of a wasp colony. This habit is known from the Icteridae and particularly from *Cacicus cela,* a species living chiefly in the northern part of South America and in Panama. Sometimes these birds build their pendulous nests so close to a wasps' nest that the dwellings of birds and insects touch each other when they sway in the wind. And yet there appears to be no other kind of "friction," though the wasps normally attack any animal that ventures too close (see p. 218).

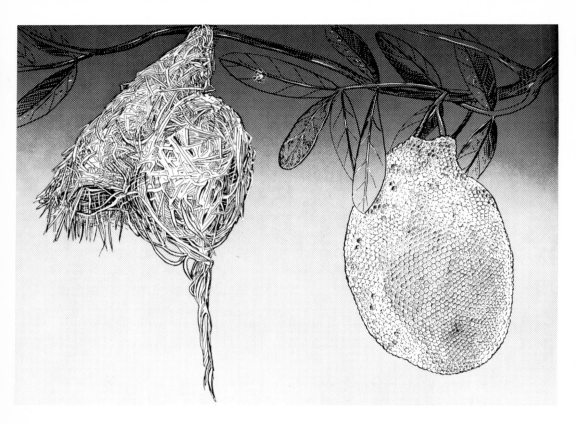

Fig. 82. Nest of a warbler from the South Pacific (Gerygone) immediately adjacent to a wasps' nest. The wasps had to be smoked out before the photograph could be taken from which this drawing is made.

Hole-breeders

Nests with their own roofs, such as those built by wrens, penduline titmice, weaverbirds, and others, may be compared with small cave dwellings in that they offer better shelter and more warmth than shallow depressions or cup-nests that are open to the sky. But the building of durable covered nests is by no means easy, and hence it is not surprising that many species prefer to raise their broods in the cavities of decaying trees, of which there is a plentiful supply in natural forests. Woodpeckers (Picidae) are among the best known hole-breeders.

Woodpeckers. Woodpeckers may be called the carpenters among birds because they are specially equipped for woodworking. Their strong claws and stiff tail feathers, which they use as supports, enable them to find a secure foothold on vertical stems and to move up and down with ease. Using their strong beaks—their most important tool—they open the galleries of insect larvae, and with their long tongues, which are barbed at the end like har-

218

poons, they pull out succulent grubs from their hiding places. Occasionally, they eat hazelnuts and conifer seeds, an important source of nourishment during winter months in temperate zones, as well as other vegetable food. For the opening of nuts and the extraction of seeds from conifer cones, they use special workshops, known as woodpeckers' anvils. A crevice in rough tree bark or a crack in a piece of wood may serve as a vise to hold a nut. With larger objects, such as spruce cones, the woodpeckers themselves prepare holes of suitable size for wedging, to enable them to extract the seeds from the cone scales by skillful thrusts with their beaks. Effective "anvils" are used year after year, and sometimes the debris of thousands of cones on the forest floor may draw the attention of an observer to a workshop above.

As ready-made dwellings may not always be available in well-managed forests, woodpeckers are often compelled to chisel their own breeding cavities out of the wood of sound trees. They are strong enough to cope with wood as hard as that of a sound beech tree. In their carpentering, they use their beaks both for chiseling away the wood and for removing wood chips from the interior. This is done chiefly by the male, and it is hard work. Hence, a cavity is often used for many years in succession. The bones in a woodpecker's skull are specially constructed and strengthened so that the strenuous hammering cannot injure the brain.

A woodpecker's beak is at the same time an instrument of communication. Whereas other birds sing to lay claim to a territory, to attract a female, or to express a mood, a woodpecker does this by using a dry branch as a musical instrument. Sitting on a selected limb, he hits it with his beak to produce a sound which travels far because of the resonance of the wood. The duration and rhythmic intervals of the drumming sequences allow these birds to recognize their own and other species. Drumming is a way of staking a claim to a territory, and a means of communication regarding the choice of a breeding place and other family matters. Where a suitable dry branch cannot be found, the woodpeckers will drum against a tree trunk or any other object with a good resonance.

When I was young, I had a great spotted woodpecker which was quite tame. He was allowed to fly about freely (pl. 94, p. 233). In summer he lived in the woods bordering our country place, but he always came when we had our lunch in front of the house. He would perch on a wooden drainpipe and announce his presence with a lively drum roll before flying down to our table to demand his daily ration of mealworms.

Woodpeckers do not build nests in the strict sense of the word. A few wood chips left in the cavity suffice to cushion the eggs that are laid on the floor. No one knew how they behaved inside their dark holes until Heinz Sielmann succeeded not only in observing them, but even in filming their intimate family life. He managed this with the largest and shyest of European woodpeckers, the black woodpecker (*Dryocopus martius*), though all specialists had told him it would be impossible.

The nesting cavity in question was twelve meters above the ground in an old beech tree. A well-camouflaged "hide" was built on the side of the trunk opposite to the flight hole to accommodate the observer and his camera. From this safe vantage point he was able to open the back of the hole with the aid of electric drills and saws. Starting from the top, he removed the back wall very gradually, and replaced it with a pane of glass. Extreme care was necessary, and each major disturbance had to be followed by a lengthy interval to allow the birds to calm down and adjust to the change. The inserted window allowed the interior not only to be watched but to be filmed as well. The birds even tolerated the glare of searchlights. They would hardly have done so if they had not been more or less tied to the spot by the fact that they were breeding. Under these circumstances, they carried on normally, watched by Sielmann and recorded by his camera.

Because of his diligence we know that the parents take turns in sitting on the eggs and in feeding the young. A parent arriving with food climbs down inside the cavity, head first. The black woodpecker carries the food in its crop and regurgitates it into his beak for feeding the young. These do not seem to notice anything until the parent touches the base of their beaks where there are sensitive skin swellings acting as releasers. As soon as this spot is touched, the young stretch their necks, open their beaks wide, and wait for the food to be pushed down their throats (pl. 95 a and b, p. 233). After each feeding, the young are warmed briefly. Should the other partner arrive in the meantime with more food, it announces its presence by a knocking signal and is answered by the bird inside. As the young grow larger, they take their food at the nest entrance.

Young woodpeckers leave their holes after three or four weeks, development depending on the size of the species. Soon thereafter, the families disperse. Occasionally, the young are chased away by their parents after only eight days and must start looking for territories of their own. When brooding time is over, woodpeckers use their old holes as sleeping places, or they build special sleeping holes.

The yellow-shafted flicker (*Colaptes auratus*) is a popular bird in North America. Like its European cousin, the green woodpecker, it likes to hop about on the ground looking for ants, though it also feeds on fruit and other vegetable matter. These birds are very conspicuous at mating time when they execute a kind of dance, using vigorous movements of their heads, and display the brilliant yellow feathers of their tails and wings. This behavior is common to both sexes and appears to be designed both for the defense of the territory and the courting of a mate. The cavity-building and brood-tending behavior of the flicker is similar to that of the black woodpecker.

Another popular bird dwelling in the eastern part of North America is the redheaded woodpecker (*Malanerpes erythrocephalus*). It has the habit, unusual among birds, of storing food for the winter in well-protected caches. It collects acorns, beechnuts, nuts, and other fruit, and stores them in cavities that it has found or made in tree trunks or fence posts. To protect his hoard from being taken by others, the bird pushes it deep into the hole and closes the opening with pieces of bark and chips of wood. This woodpecker possesses not only a hole-dwelling in spring, but a number of small well-camouflaged cavities for its stores as well.

Hornbills. The hornbills (Bucerotidae) are hole-breeders of an entirely different kind. These bizarre-looking creatures live in Africa, south of the Sahara, and in warm parts of Asia (as far east as New Guinea). One cannot help wondering about the significance of their monstrously developed beaks, and may even feel sorry for them because of the heavy load they carry on the front part of their heads (fig. 83). However, this is not so bad as it looks. The horny substance is very light and so is the bone structure of the beak which, in common with bird bones in general, contains many air-filled chambers. The extension of the head by gigantic beaks gives these birds a considerable advantage. Their chief food consists of fruits of tropical trees, which usually dangle on the tips of thin branches. If their beaks were short, birds of this size and weight could not reach the fruit because they could not get sufficiently far out on the branch.

These birds have aroused compassion by yet another peculiarity. During breeding time, the female lives for many weeks walled up in a cavity in a hollow tree, the entrance of which is almost completely closed by a mud wall. Only a narrow slit remains through which the female can receive food from the male (fig. 83). But as so often happens, the facts are rather different from one's original impression. The female is not "incar-

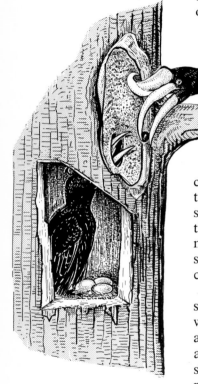

cerated"; she builds the confining wall herself from material provided by the male. He brings her lumps of moist soil which occasionally he has wetted with his saliva; to this she adds excreta and food residues and finally hammers the curious mixture into place with vibrating lateral strokes of her broad beak. When the mass dries, it becomes very hard.

Her close confinement means that she can devote herself to the matter of incubation and tending of the brood without disturbance. Moreover, she uses this period for a rapid complete molt, another unusual peculiarity. As all her wing and tail feathers drop off simultaneously, she would be unable to fly in any case. By way of compensation, her new feathers also grow all at the same time, so that after a few weeks she boasts a completely new plumage. This happens none too soon, since in the meantime the young have hatched and, after passing through their first nestling stage, have developed such appetites that the father alone can no longer feed them. The mother now breaks down the wall and helps her partner to collect food for the young. No sooner does she open the entrance than the young birds start renewing

*Fig. 83. Immured in a tree cavity, a female hornbill (*Buceros rhinoceros*) receives food from the male through a narrow slit. The female and eggs in tree shown diagrammatically.*

222

the wall; they voluntarily remain in their safely confined space for several more weeks, until they finally feel the urge to be free and break down the wall they themselves have built. Sometimes the family is at cross-purposes about this. For as the birds did not all hatch at the same time, the older birds may be anxious to open their prison while the younger ones are not yet ready and try to repair any breach in the wall as soon as it is made.

There are forty-five different species of hornbills. Some are no bigger than crows, while others are the size (but not the weight) of turkeys. Among the large species are the ground hornbills (*Bucorvus*), two species of which inhabit the savannas and dry regions of Africa. They differ from other hornbills by seeking their food on the ground. Snakes, grasshoppers, and other prey are easy to come by, but discovering tree cavities of sufficient size is more of a problem. They find suitable holes only in baobab trees; therefore the distribution of these birds coincides more or less with that of this remarkable savanna tree. The *Bucorvus* species are the only hornbills that do not wall up the nesting cavity. Though in these species, as in all others, it is the female alone that sits on the eggs, relying on her partner to feed her, the female *Bucorvus* does get up from time to time for short walks or hunting expeditions. And because physiological functions of animals tend to be adapted to their modes of life, she does not molt abruptly, but changes her plumage gradually as most other birds do. The young hatch after one month and stay on in the brood cavity for three months. Their parents feed them for another nine months, and they take a further two years to become sexually mature. The parents are mated for life, which is rather unusual among birds, and lead an exemplary family life with their offspring.

In addition to woodpeckers and hornbills, many other birds, including most parrots and many owls, breed in holes or cavities. But not all of them nest in hollow trees. Some inhabitants of open and treeless country, and also birds in other regions, excavate passages and subterranean brood chambers in the ground. This is a useful art, widespread among insects and other arthropods as we have seen earlier, and also among mammals, as we shall see later.

Kingfishers. One well-known species that nests in soil cavities is the European kingfisher (*Alcedo atthis*).

Bird lovers enjoy watching this small bird with its splendid iridescent plumage as it darts across the water or perches on some pole on the bank in wait for prey. It looks rather like a tropical bird, and it is true that its family (Alcedinidae), which is closely related to the hornbills, belongs chiefly to the tropics and subtropics, where there are over eighty different species. *Alcedo atthis,* the only European species, is also found in Asia and North Africa. The six American species are very similar in their digging and breeding habits to the European kingfisher that will be described here. Its food consists of small aquatic insects and fish. When it has spotted a fish from its perch, it dives headlong into the water, catches it with its long beak, then hits it against the ground or a post to kill it, and swallows it head first. But when a male bird offers a fish to his intended mate in a courtship ceremony or when he feeds his young, he holds it in his beak the other way round to enable the recipient to swallow it headfirst. Such good manners are only practiced during the mating and breeding season. At other times, the kingfisher is an unsociable fellow and tolerates no other bird of his own species in his territory.

The kingfishers' nesting sites are on the banks of streams or lakes, or on other steep slopes not necessarily in the immediate vicinity of water. Both partners share the work of building, brooding, and tending the young. They start by loosening the soil of the bank with their beaks to make a hole. When this is sufficiently deep, they use their legs as well, vigorously throwing out the excavated material. They dig a slightly ascending passage half a meter to one meter in length. The end is widened into a brooding chamber. Though they do not collect nesting material, the cavity is nicely padded. The kingfishers rid themselves of the bones and scales of the fish they have eaten by regurgitation. This half-digested bony residue, which is soft and friable like cigar ash, forms a cushion for the eggs and—after about three weeks of incubation —for the young birds. Feeding time is a very orderly affair. When the parent bird arrives, it darkens the flight tube, and the shadow acts as releaser for the opening of the beak in the little bird nearest to the entrance. When it has been fed, the young all move up one place in a circle, so that each young bird gets its fair share.

The young kingfishers squirt their thin excrement into the entrance tube. As this tube slopes down toward the outside, the excrement trickles down, coating the tube.

The passage is never cleaned and it is easy to imagine what it must smell like. The parent birds have to fly in and out through the stench and at this time seem to feel a great need for bathing. Their diving into the water when they catch fish does not satisfy them: they can often be seen taking baths and splashing about to clean themselves properly. The young birds are fledged three or four weeks after hatching, and the family disperses soon after.

Birds that build clay shelters. The true ovenbirds (Furnariidae) are natives of tropical Central and South America.* Their many species inhabit lowland forests, sea coasts, or mountainous regions. Their nests are as varied as their habitats; they may be cup-shaped or spherical, situated in rock cavities, in hollow trees, or in holes in the ground. Of their two hundred species, six have attracted so much attention by their building activities that the whole family has been named "ovenbirds" after their highly original structures. They do not try to find a cavity but build one out of clay. In this way they have a hole to breed in even where natural holes do not exist. Though their nests are larger and less dainty, their style of architecture is reminiscent of that of the potter wasps (*Eumenes*) that we encountered in an earlier section. The name "ovenbird" is very apt (the Latin name *"Furnarius"* means the same), because their conspicuous nests do indeed look like little bakers' ovens. These ovens may be erected on branches, on telegraph poles, on roof tops, or, in pasture land, on fence posts. The birds themselves also like to perch on similar high points, where they can scan their territory. They tolerate no rivals near them.

Both partners cooperate in the building of the nest. Dry clay does not interest them. Their building instinct is only awakened through rain and the sight of mud. Under favorable circumstances a nest may be completed in two weeks. The work involved is considerable because the birds carry about two thousand little lumps of clay to the building site and deploy them skillfully with their beaks and feet. Vegetable debris, straw, and cow dung or other feces are mixed with the clay to give greater cohesion to the walls. The plan of the structure (fig. 84) ensures excellent protection for eggs and brood. The side walls rest

* The North American bird called ovenbird (*Selurus aurocapillus*) is not a member of this family, but a member of the wood warbler family (Parulidae).

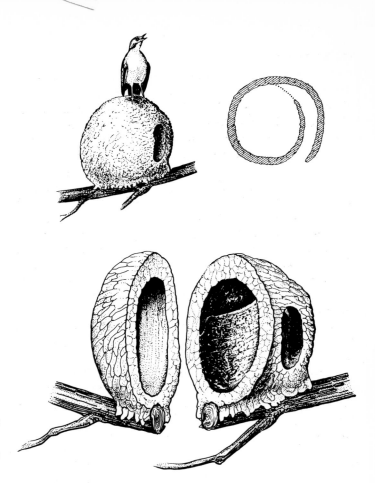

on a pediment of clay and are covered with a vaulted roof. Its horizontal cross section has a slightly oval shape. At first they leave a hole, about ten centimeters in diameter, in one of the longer walls. Then they build a curved partition, starting at one of the edges and leading inward, which divides the cavity into a narrow antechamber and a spacious brooding area (fig. 84, bottom). A gap just large enough for the birds to slip through is left between the top of the partition and the roof of the nest. Then the brooding chamber is padded with fine grass. Some five to six weeks after the eggs are laid, the young birds are sufficiently fledged and soon leave the nest. They never return to it because the dry ball of clay, while affording protection against enemies, is turned by the tropical sun of early summer into a veritable baker's oven in which it would be unbearable to stay.

The tailorbird

We have now met quite a number of craftsmen among the birds: basket-makers, weavers, carpenters, and potters, all skillfully plying their trades. But it does sound incredible that a bird should fashion its nest with needle and thread. The tailorbird (*Orthotomus sutorius*), a close relative of our warblers, lives in southern China, India, and Southeast Asia. It is quite tame and often nests in gardens, orchards, hedges, and shrubberies. It begins its activities by selecting a large leaf. Without detaching it, the bird twists it with its beak and legs into the shape of a bag and sews the edges together (fig. 85). For a needle it uses its long pointed beak; for sewing thread it chooses spiders' silk, bark fibers, or short cotton fibers which it cleverly twists into longer threads. Or it may take advantage of the products of civilization and pick up a discarded length of string. The bird proceeds by making holes in the overlapping edges of the leaf and pulling through its thread, which it prevents from slipping out again by tying a knot. Twisted cotton fibers often stay in place without this expedient because the ends are splayed and tufted. Occasionally two or more adjoining leaves are sewn together. The way the bird coordinates its beak and legs in this intricate work is quite amazing. Finally, a cozy nest made of sheep's wool, plant down, and the like is placed inside the completed bag, and in this green nursery, which is practically invisible from outside, the eggs are laid and the little birds grow up.

Various other birds more or less closely related to the tailorbird employ similar but more rudimentary sewing techniques. The fan-tailed warbler (*Cisticola juncidis*), a songbird of the family Muscicapidae which occurs in southern Europe, builds a deep, bag-shaped nest of plaited grass and fibers. To strengthen its structure it occasionally bores holes into the narrow leaves of grasses with its beak and pulls spiders' silk through them. An eastern South African species of the same genus hides its nest under leaves and sometimes sews them together in the manner of the tailorbird. But no other species masters the art of sewing to the same degree.

Edible nests of swiftlets

The tailorbird is a specialist in the art of handling its nesting material. Other birds specialize in the way they *produce* theirs. The swiftlets are unique in this respect.

Fig. 85. Tailorbird with upward pointing tail perched beside its skillfully sewn nest.

They make their nests of hardened saliva. Strangely enough, the Chinese love to eat them. These nests have been collected and sold by the thousands in Chinese markets for centuries. Nobody knew what they were made of, how they were made, or whether they had any nutritional value. Today it is known that they have practically none, but people do not seem to mind. As prepared by Chinese cooks with rare ingredients and condiments, they are considered a special delicacy.

However, we are not here concerned with cookery. We are interested in the kind of birds the swiftlets are and the way they build their nests.

The swiftlets belong to the family of swifts (Apodidae). The common swift (*Apus apus*) is the best-known European representative. It looks like a swallow, and many take it to be one. But great similarities in appearance and behavior notwithstanding, the swifts are not swallows but close relations to the hummingbirds. They can fly better than any other bird, and their adaptation to life in the air is so extreme that they are hardly able to move on the ground. Their scientific name *Apus* is Greek for "without feet." This is an exaggeration, but their feet are small and hidden in their plumage. When these birds are at rest, they cling to branches or vertical walls with pointed claws. But they rest very little. Most of the time they rush through the air on sickle-shaped wings, hunting small insects, and uttering their unmistakable shrill cry, *Sreeeth!* They even catch their nesting material in flight, including seed hairs, flower petals and other small plant parts carried high by the wind, as well as dry grass, feathers, scraps of paper, and other particles. From these materials they build shallow nests in tree cavities, rock crevices, or corners in buildings, using saliva to cement the small particles together. This use of saliva as a quick-drying binder in nest-building is a peculiarity of the whole swift family. The common swifts appear in central Europe at the beginning of May. They stay only three months and return to their African winter quarters as soon as breeding is over.

The family of swifts, which comprises many genera and over eighty species all very similar in appearance and habits, is distributed over a large part of the globe. Of the seventeen species of swiftlets (*Collocalia*) inhabiting southern Asia and Indonesia, only a few build the famous edible nests. Very large colonies of these birds often nest together in rock caves or on overhanging cliff faces. Both parents take part in nest-building and in the care of the brood. As breeding time approaches, their salivary glands swell to an enormous size.

When the birds start building, they fly repeatedly (ten to twenty times) toward the chosen spot on the rock face, never moving away more than a few meters. Each time they touch the wall, they place some of their saliva, which is a viscid, thread-forming, quick-drying fluid, onto the stone with their tongues. Tracing first a semicircle, which forms the outline of the future nest, they build this up into a raised rim and gradually, layer by layer, form a console-shaped nest. A coating of saliva covering the

Fig. 86. The nests of swiftlets, made from their hardened saliva, are considered a delicacy to eat.

stone at the back gives strength and cohesion to the little structure, which has the appearance of a white, translucent cup. The eggs are laid in this cup without padding of any kind (fig. 86), and the young birds have to make do with an unlined nursery. Their parents feed them little food balls consisting of hundreds of small insects stuck together with saliva, a substance they seem to use for everything.

Good things take their time, and nests made of pure saliva no doubt have a long history of phylogenetic evolution. The majority of swiftlets still represent earlier stages of this development and mix plant parts, moss, or feathers into the secretions of their salivary glands. Such nests are, of course, of considerably less interest, if not quite useless, from the culinary point of view. But for us this is hardly a reason for not looking at the nest structures of some other swift species.

Whereas the swiftlets build their nests on rocks, the palm swifts, as their name indicates, attach theirs to the swaying leaves of palm trees. Three species of palm swifts inhabit America, and one (*Cypsiurus parvus*) represents the group in Africa south of the Sahara, Madagascar, India, and the Philippines. Like the common swift, they use as their nesting material cotton fibers, plant hairs, or feathers of other birds which they have caught in flight and cemented together with their saliva. But whereas the nests of the American species are bag-shaped and give excellent protection to their eggs and young, the nest of the small *Cypsiurus* has a strange form and appears to be awkwardly constructed. It has the shape of a tiny shallow spoon, and as the leaves of palm trees tend to hang down, it looks as if the eggs are bound to fall out. This does not happen, however, because the bird glues them to the bottom of the little shallow dish with saliva, the universal family nostrum. As soon as the young birds hatch they attach themselves to the nest with their claws and spend most of their nestling phase in an upright position.

The lesser swallow-tailed swift (*Panyptila cavennensis*) has its home in Central and South America. Its nest material is similar to that of the palm swift. And yet there is the greatest contrast imaginable between the precarious position of eggs and young in the nest of *Cypsiurus* and the sheltered condition of the breeding cavity constructed by *Panyptila*. This bird takes great trouble with its building activities and is supposed to build for about half a year. The nest may be suspended from an overhanging rock face, as is shown in figure 87, or it may hang from a tree. The brooding area is in its upper part. The bottom part ends in a hanging tube, sometimes over sixty centimeters long, into which the bird may be seen rushing from below in a flight of amazing speed and agility. Halfway up there appears to be another entrance, but this is a blind alley. The only plausible explanation for this structure seems to be that the accessible and conspicuous mock entrance serves to deflect intruders from the true entrance to the nest.

A living nest with central heating

It is a long way from the hot tropics to the icy wastes of the Antarctic with its months of darkness and raging blizzards. Can birds brood under such conditions? The emperor penguin can and does. Of course, it could never hope to find any warm and soft nesting material, but it does not need to because nature has equipped it with a warm nest attached to its body. In some way this strange story links up with the first chapter of the book when we looked at the architectonic structure of animals' bodies. Let us give it a closer look for better understanding.

As swifts are adapted to the point of extremity to life in the air, penguins (Spheniscidae) are adapted to life in the sea. They dive for fish, squids, and crustaceans. With the aid of their wings, which have been transformed into flippers, they swim with an agility reminiscent of dolphins. They use their feet as rudders, and this is the reason why their feet are set at the very end of their bodies. This, in turn, causes them to adopt an upright posture on land in order not to fall over, and often makes them look amazingly like caricatures of ourselves. They use their stiff tail feathers to support this upright stance as woodpeckers do when sitting on a vertical tree trunk. Although the penguins look clumsy on land, they can walk and hop quite well and slide across the ice on their bellies.

They live in the southern hemisphere and most of them inhabit temperate zones. The emperor penguin (*Aptenodytes forsteri*) is the largest species. It measures about one meter in the upright position and weighs as much as forty kilograms because of a thick layer of blubber under its skin which enables it to withstand the rigors of an antarctic winter, where temperatures may drop to about $-60°$ C. ($-112°$ F.) and icy blizzards rage for days.

These conditions make it even more strange that in March, of all seasons, when autumn starts in the Antarc-

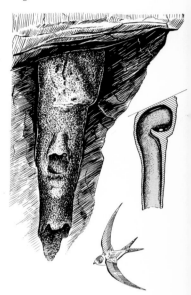

Fig. 87. Nest of a lesser swallow-tailed swift under an overhanging rock. Entrance at the bottom. In middle of front, a mock entrance that ends blind. The cross section (right) shows brooding area in top part. Flight tube is about 60 cm. long.

tic and winter looms ahead, the emperor penguins leave the sea where they find plenty of food. Prepared for the rigors to come by a thick layer of fat, they trek southward across the ice. After weeks of traveling on foot across the frozen ocean, they finally reach dry land. There they often form large colonies and court and mate undeterred by the inclement season. In May or June, in the depth of the antarctic winter, the female penguin lays a single egg and at once entrusts it to her male partner. He takes it onto his webbed feet, which are richly supplied with blood vessels and therefore quite warm, and covers it from above with a brooding pouch, a deep fold that hangs down from the abdomen of both sexes (fig. 88). This is the living nest in which the egg finds shelter and warmth. It fits into this pocket so snugly that it does not fall out even when the father, who incubates it for two months, walks about. But usually he stays upright in one place; he has to economize his energy, since he does not eat at all. When snowstorms rage, the fathers huddle together to keep each other warm.

During all this time, the mother makes no use of her brood pouch. As soon as the egg is handed over, she returns to the sea to eat her fill and to recuperate. When she has regained her original strength and her former weight, she returns to the breeding place, taking with her in her stomach a load of fish weighing several kilograms, since she will find no food there. As soon as she arrives, she searches for her mate. Pairs are able to recognize each other by the sound of their voices. No time must be wasted because the father, who has been fasting for three months, has lost a third or a half of his original body weight and the chick has hatched. The family enjoys a wholesome meal of predigested fish from the mother's stomach, and starvation is at end. The mother now takes over the care of the chick, which is no less safe and snug on her feet and in her brood pouch than with its father. Relieved from duty, he, in turn, migrates to the sea where he recuperates for two to three weeks before returning to his family with plenty of food in his stomach.

The young penguin stays in its mother's brood pouch for about five weeks. Later, all the young birds sit together in a large group, a kind of penguin kindergarten, where they are looked after and fed by their parents. Their growth is slow because the food which the parents have brought must suffice for the whole family. It takes five months for the young to become independent. With

Fig. 88. A male emperor penguin incubates an egg. Despite the arctic cold, he keeps it warm, nestling it between his webbed feet, which are rich in blood vessels, and covering it with his brood pouch.

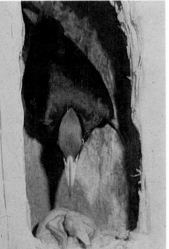

Plate 94. *A female of the great spotted woodpecker* (Dendrocopus major) *with food for the young at the entrance of the nesting hole. (See p. 219.)*

Plate 95 a. *A breeding cavity of a black woodpecker in an old beech tree. While birds were brooding, a window was made in tree and a pane of glass inserted. Through it, events in the nest could be filmed. Here, the parent bird arrives with its crop filled. The young do not react until touched on sensitive skin swellings at the base of their beaks.*

Plate 95 b. *Young woodpeckers stretch their necks and wait for the food to be pushed down their throats. (See p. 220.)*

233

Plate 96. *A great bowerbird has placed small bones and snail shells in front of his bower. He holds a red ornament in his beak as a prized possession. (See p. 240.)*

Plate 97 (right, top). *Male satin bowerbird building his bower. (See p. 238.)*

Plate 98 (right, bottom). *A Lauterbach bowerbird has used blue berries for decoration. The red berry he holds in his beak is evidently special. With it he tries to attract a female. (See p. 242.)*

Plate 99. Bower of the orange-crested gardener in the rain forest of New Guinea. The two openings in front of the hut are connected inside by a semicircular passage. The bird has covered column between the two openings with dark moss. It is decorated on one side with blue iridescent beetles, in the middle with yellow flowers, and on the other side with broken shells. In front of the bower is a fence plaited from twigs and decorated with brightly colored fruits (sometimes with flowers as well), which forms boundary of the "garden." The male (left) has just rushed out of tunnel and greets the female by displaying his nuchal crest. (See p. 243.)

the approach of the antarctic summer, the whole colony migrates back to the coast. The breeding grounds are once more deserted.

No one knows what induced these strange birds to choose such an inhospitable area and such a cold climate for raising their brood, or to accept the hardships and dangers connected with this choice. But their success shows that they have made good use of an extraordinary niche. Well-equipped to withstand the rigors of prolonged cold, they migrate to an area where they may be expected to be less disturbed than anywhere else on this globe and where they have no enemies because no one else can survive there. Later in the year, when the food demands of the growing offspring become difficult to meet, but when the young birds have reached a stage in which they can fend for themselves, the situation is changed. The pack ice has melted; the way to the open sea is much reduced, and the conditions are favorable for the young penguins to start out on a life of their own.

Bowerbirds and their bowers

During courtship, many birds display magnificent ornamental feathers; others try to win a mate by melodious song or flight aerobatics, and still others by the building of nests of outstanding workmanship. We have seen this with some weaverbirds, where the male reaches the consummation of his desires if he can get a female to accept a nest he has built as a place to brood in.

Courtship behavior of bowerbirds (Ptilonorhynchidae) is in a class by itself. In this unique group, the male constructs a special bower, decorates it, and dances before it, with the sole object of enticing a female to enter it and mate with him. The bower is a structure designed exclusively for love-making, not for brooding. The actual nest for the brood is built by the female long after the mating ceremony. It is a simple, cup-shaped affair built in the branches of a tree sometimes a few hundred meters distant from the bower which always stands on the forest floor. The female alone incubates the eggs and tends the brood while the male continues to occupy himself with his bower, presumably in the hope that he may entice another female.

Sixteen different species of these remarkable birds are known, eight from Australia and eight from New Guinea. They are closely related to the birds of paradise but lack the gorgeous plumage their cousins display before the eyes of their females. The bowerbirds offer something entirely different. The structures they build vary greatly from species to species, and there is also considerable variety in the choice of ornamentation and in the courtship ceremonies themselves. Most of these birds live in isolated areas and are, moreover, very shy, so that they are difficult to watch. Some of them are so rare or so sensitive to the slightest disturbance that hardly anything is known about their behavior. But it has been possible to penetrate the secrets of some species. I shall discuss here only a few examples.

The satin bowerbird (*Ptilonorhynchus violaceus*) lives in the rain forests of eastern Australia. The male, which is about the size of a pigeon, starts building his bower long before the onset of the mating season, and chooses for it a not too shaded spot on the forest floor. First, he clears all debris from an area of ground about one meter square. Next, he makes a kind of avenue by sticking bunches of straight twigs, twenty to thirty centimeters long, into the ground in two parallel rows (pl. 97, p. 235). At the southern end, where most light will penetrate in the course of a day, he prepares a dancing floor, putting down a carpet of fine twigs and grass. He then collects bright objects to decorate it. He prefers dark blue and yellow-green colors, possibly because the blue ones match the bluish-violet radiance of his plumage, and the yellow-green ones resemble both the yellow-green color of his beak when fully mature and the yellow-green plumage of the female. He decorates the bower with blue and yellow flowers, blue-colored berries, and parrots' feathers. Where he is near human habitations, he adds to his display such products of our civilization as glass beads and strands of colored wool and tinsel. But this is by no means all. He actually paints the inside of his bower with the juice of the blue berries that he crushes in his beak. Sometimes he even uses a tool for the purpose. He picks up a piece of fibrous bark simultaneously with a squashed berry and proceeds to use the bark, soaked in the juice of the berry, like a brush or sponge.

Though the bower itself is finished in a few days, the bird's work does not end there. He constantly removes withered flowers and dried-up berries, replaces them with fresh ones, and generally adds to his collection as much as he can. In this he is completely unscrupulous: he steals from neighboring bowers if their owners happen to be away. When some pieces of blue-colored glass were placed on a suitable spot by an ornithologist, they were eagerly carried away, and soon they appeared in all the bowers of the district. As the pieces had each been marked with a number, it was not too difficult to trace their migrations from one bower to another as a result of constant multilateral thieving. Maintenance, too, is a laborious task. The heavy tropical rains damage the bowers, wash off the paint, and necessitate repairs. Preparations may thus keep a bird busy over many weeks.

When mating time approaches, the cock redoubles his efforts to attract the attention of a female to his bower and display by intoning a raucous courtship song. Bowerbirds do not excel as singers. But some species adopt a highly personal form of acoustic wooing by imitating other sounds with amazing fidelity, be it the song of other birds or the roll of thunder.

If a female has come to his bower, the male gets more and more excited as he tries to win her. He hops around the bower, dances on the place in front, and constantly picks up various ornaments with his beak to show them to his visitor. His excitement enables him to produce a very special effect: he can change the pale blue of his iris into a dark, bluish-violet hue matching his plumage, a color which is so markedly preferred in his exterior and interior decoration. Courtship play continues until at last the female shows her willingness, slips into the arbor, and mates with him.

The splendid dark-blue plumage of the males does not develop until their fourth year. In the first years of their lives, the males are the same pale yellow-green as the females, even though they may be sexually mature at a younger age. The urge to build also starts quite early, and though first bowers tend to be far from perfect, they may nevertheless be successful. On one of his filming expeditions in Australia, Heinz Sielmann observed a female that was courted by an old male slip into the bower of a young green bird. The old male noticed it too and took his revenge. He destroyed the bower of his rival and carried away his treasures. The young male did not offer any resistance. He respected the seniority of the other bird proclaimed by his dark blue plumage. The female gave in, too, and entered the bower of the older suitor.

The great bowerbird (*Chlamydera nuchalis*) inhabits the tropical bush of northern Australia. It is the largest of its group, and its bower is especially elaborate and impressive. The two rows constructed from grasses, sticks, and thin twigs are inclined toward each other and form an arbor about forty centimeters high and one meter long, closed at the top and open on two ends (pl. 96, p. 234). The dance floor lies in front of one of the entrances. This species chooses mainly white or pale yellow objects for its decoration. Sielmann counted five hundred bleached kangaroo bones and over three hundred pale yellow snail shells in front of one bower. Even greater feats of collecting have been reported. Near human settlements, the bird is attracted by man-made objects that glitter in the sun. Bottle tops, metal buttons, hair curlers, nails, and other things have been discovered in front of bowers. A shrewd farmer retrieved some teaspoons and even two missing car keys by searching the bowers of the neighborhood. The birds seem to play with these objects, arranging them in groups and changing the arrangement over and over again. One such bird had got hold of a large but light-weight tin mug which it proudly showed off as its most highly prized possession (fig. 89). At the peak of excitement the males surprise the females by displaying a nuchal crest of red feathers which is normally hidden under their inconspicuous plumage. Sielmann observed and described a visit of a female and the display of the male whose showmanship was indeed remarkable.

"Carrying a snail shell in his beak, he tried to entice the female into the love nest. While she settled down in the house and watched from there, the male began his dance. Observing him from her front seat in the stalls, the lady seemed to weigh up what he had to offer her in the way of wealth and personal appearance. He showed her the most beautiful pieces of his collection, holding them high in his beak, but she did not seem to be impressed by any of them. She appeared more interested in his physical beauty. He displayed his best points, lifted his wings and hopped enticingly round the bower. When finally he ruffled his neck feathers and displayed the shining red collar which had not been visible before, she gave in. She allowed him to enter the bower and soon they were mated." *

Another species of the same genus, called *Chlamydera Lauterbachi* after its German discoverer, lives on the tropical island of New Guinea. This bird, too, builds an avenue, but in addition to the two side walls of the basic structure, it places similar rows of twigs across both

---

* Heinz Sielmann, *Lockende Wildnis—Das Reich der Drachen und Zaubervögel* (Gütersloh and Vienna: Bertelsmann Sachbuchverlag, 1970), pp. 127 f.

Fig. 89. The great bowerbird has decorated his dancing floor mostly with snail shells. He proudly holds up an unusual ornament—a tin mug.

openings, thus creating an enclosed space like a cloister which it decorates with various ornaments. A diagram illustrating the bowers of the Lauterbach bowerbird and the satin bowerbird is shown in figure 90. The strange structures of Lauterbach bowerbirds with their displays of groups of blue berries, small blue pebbles, and some red berries had long been known, but no one had succeeded in seeing the courtship displays of this very shy species, let alone filming them.

When Sielmann on his arrival in the highlands of New Guinea announced this very purpose, he met with universal skepticism. It would not be too difficult to find the bowers in the dense bush, but the birds always kept away. However, helpers from the island built him a well-camouflaged hide close to a bower, advantageously

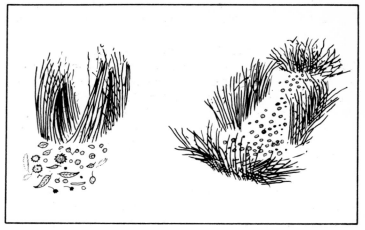

Fig. 90. Diagrammatic represen-
tation of bowers: left, of the satin
bowerbird; right, of a Lauterbach
bowerbird.

placed for filming. After a period of long and anxious waiting, he succeeded in observing the courtship dance of this bowerbird for the first time and managed to film it. As soon as a female approached, the male picked up a red berry in his beak and made inviting nodding movements with its head (pl. 98, p. 235). There were far more blue berries than red ones in the bower, but the red color appeared to be more impressive. It is important that this display with berries should not be looked upon as an offer of food. It has never been observed, either with this or with any other species, that the courted female has eaten any berries. Like the bleached bones of kangaroos, like flowers and other ornaments, they are simply a means of signaling to a female that a male is waiting for her in the bower.

The measures taken by the male to attract the attention of a female and the courtship displays designed to win her favor are probably most highly developed in another group of this fascinating family. I choose the orange-crested gardener (*Amblyornis subalaris*) as my example. It is a bird the size of a starling, with inconspicuous plumage, which lives in the dark, inaccessible mountain forests of New Guinea. At mating time, the male builds a little hut on the forest floor. I use the word "hut" advisedly, because it really is a hut with a rainproof roof, with a circular passage inside designed for a tryst, and a colorful mosaic between the two openings. In front of this structure, there is a well-tended garden strewn with flowers and separated from the surrounding area by a fence richly decorated with yellow and red fruit (pl. 99, p. 236). A. P. Goodwin, the first author to report on their bowers, described these creations as the most beautiful objects ever constructed by birds. His view remains valid today.

As with the Lauterbach bowerbirds, the bowers were known, but all efforts to watch the birds at work or observe their courtship displays remained fruitless for a long time. Even Sielmann nearly failed, despite his determination. With the aid of experienced local helpers, he finally managed to penetrate to the courtship area. There, in a cleverly constructed hide, he and his assistants waited endlessly with their cameras for a chance to film. But under a constantly overcast sky, the light in the dark rain forest was insufficient for his purpose. At long last, after weeks of waiting in the hot and clammy hide, their patience was rewarded. An occasional ray of sun penetrated the dense canopy, allowing them to record on film events that seemed almost incredible.

When the party had first arrived, the hut of the orange-crested gardener had already been completed. But heavy downpours played havoc with it, and they could repeatedly watch the bird restoring order to house and garden. The dome-shaped roof of the hut, made of densely interwoven twigs, had not suffered, as it was built around the stem of a sapling for support. The front side of this stem, or center column, was covered with a thick layer of very dark green moss. It is the habit of the orange-crested gardener to place the treasures of his collection on the dark moss in the manner of a jeweler exhibiting his wares on a ground of dark velvet for better effect, and here the bird had much work to do after each heavy rainstorm. He insisted on perfect order and regularity. To the left, there would be a collection of glittering blue beetles; to the right, the shiny fragments of broken snail shells; both groups were separated from each other by a line of yellow flowers. All this was meticulously arranged and stuck into the moss at carefully chosen intervals. Sielmann himself describes the bird's decorating behavior as follows:

"Every time the bird returns from one of his collecting forays, he studies the over-all color effect. He seems to wonder how he could improve on it and at once sets out to do so. He picks up a flower in his beak, places it into the mosaic, and retreats to an optimum viewing distance. He behaves exactly like a painter critically reviewing his own canvas. He paints with flowers; that is the only way I can put it. A yellow orchid does not seem to him to be in the right place. He moves it slightly to the left and puts it between some blue flowers. With his head on one side he then contemplates the general effect once more, and seems satisfied." *

The same care is lavished on the decoration of his fence with colorful fruit.

During the many weeks of patient waiting, only two females visited the bower. When this happened, the male went inside his hut and called with a rattling sound. The female would hesitate and look alternately at the ornaments of the mosaic and those of the fence. Then the male would start to dance. It was a circular dance during which he displayed his red crest. Like a red streak of lightning he rushed from the hut, passed the mosaic, and disappeared almost at once through the second opening into the darkness of the bower. While dancing about in this lively manner, he uttered loud cries. At first, the female had seemed frightened and had jumped back, but gradually her movements followed those of her partner. Finally, both danced a *pas de deux* in which they almost touched, until, quite suddenly, they disappeared into the hut. The consummation was reached.

What passes in the mind of a bowerbird when he builds and decorates his bower?

Naturally, I cannot answer this question. No one can, because there is no means of bridging the gap to the consciousness of other living creatures. Some people are convinced that our mental states differ fundamentally from those of animals and that only we humans possess the faculty of thought. Even in displays as elaborate as those of the bowerbirds, they see but the workings of innate instincts and the effects of natural selection. Such a theory cannot be refuted. But I myself do not believe in it. The behavior of these birds has too much similarity with human behavior in comparable situations. Those who consider life on earth to be the result of a long evolutionary process will always search for the beginnings of thought processes and aesthetic feelings in animals, and I believe that significant traces can be found in the bowerbirds. Moreover, these birds, which are after all not our phylogenetic ancestors, are not the only animals exhibiting similar traits. In chimpanzees, too, not only insight into the consequences of their actions but also evidence of

---

* H. Sielmann, *Lockende Wildnis,* p. 152.

aesthetic feelings can be found. When they are given the opportunity of painting primitive pictures with color and brush, they seem to enjoy what they are doing.

On the other hand, it would be wrong to expect too much thought behind the actions of birds. They build cleverly constructed nests without having been taught and without much trial and error. Their activities are clearly directed by innate drives. In some species, the very first attempt at building is completely successful. But we have also met examples of birds that learn by experience and whose later efforts are improvements on their first, and even of birds that themselves destroy a nest that is less than perfect and replace it by a more accomplished one. In general, it is the birds that live under changing circumstances that show the greatest mental agility. Adaptation in them is never as one-sided as in certain other species. They are inquisitive by nature and capable of making use of personal experience and learning from it. Their drives leave room for a certain flexibility, and this may deceptively obscure the degree to which they are ruled by instinct and are devoid of any deeper insight into their actions.

We have already seen examples that bear this out. The mallard duck who chose to nest by a tiny ditch in the center of a big city possessed without doubt sense organs of the necessary acuity and a sense of orientation sufficiently strong to enable her to recognize the difficulties that beset her return to the river. But she lacked insight into the situation and, with a mindless courage, took a dozen ducklings with her on a march through the dangerous maze of a large city. The other duck, who built her nest on the flat roof of a high urban building from which her ducklings fell to their death on leaving the nest, did not show any higher degree of insight.

Or take the mother hummingbird who allowed another hummingbird to steal parts of her nest literally from under her body, until the nest lost its bottom and the two young nestlings she sat on fell out and perished. No doubt, she could have defended her nest, but she lacked the intelligence to recognize the effect these thefts of nesting material would have for herself and her brood.

An entirely different case is the European blackbird (a close relative of the American robin) that had built its nest in a bicycle shed 150 meters long (fig. 91 and pl. 100, p. 246), placing it upon a horizontal roof beam at the back, where this formed an angle with one of the

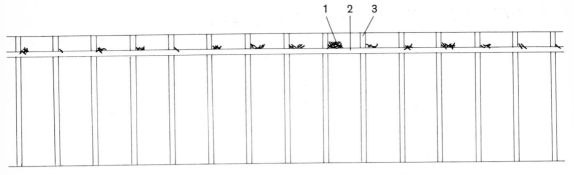

Fig. 91. Part of back wall of a bicycle shed, 150 m. long (diagrammatic sketch). (1) Nest of a blackbird, in the angle between a horizontal beam (2), and one of the roof struts. Over a length of 60 m., fourteen nests had been started in corresponding places.

roof struts. At corresponding points, there were, in all, fourteen half-completed nests; eight to the left and six to the right of the occupied nest. The bird had been muddled by the topographic similarity and had started on several different places. It doubtless was able to *see* the result of its uncoordinated activity. But in order to *act* methodically, it would have had to grasp the situation with its mind. Still, it managed in the end to finish one of the nests. Another blackbird that tried to nest on three ladders suspended horizontally on a garden wall (fig. 92) was less lucky. It started building in nine different places and took so long over it that it never achieved a completed nest.

Though these observations and many records of similar behavior suggest that not very much thought is associated with the actions of birds, it would not do to generalize. There are, without doubt, considerable differences in the mental levels of birds and their activities. Nor would it be meaningful to argue, as some scientists have done, that a bowerbird cannot have an aesthetic

Plate 100. View of the bicycle shed. It is as long again in the direction toward the viewer as the part shown here.

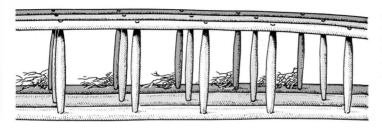

Fig. 92. Three ladders hanging horizontally on a garden wall. Abortive nest-building attempts of a blackbird appeared at nine different rungs. It never succeeded in finishing a nest.

understanding of his displays because they are triggered off by his sex hormones, and that without these there would be neither building activity nor courtship display. Do not human males, too, court women only under the influence of their sex hormones? Oriental potentates knew very well why they employed eunuchs to guard their harems. Anyone who accepts these conclusions for birds will have to consider them no less valid for humans.

MAMMALS

Mammals, like birds, maintain a constant body temperature, their fur taking the place of the birds' feathers in preventing severe heat loss. Whereas birds lay eggs and incubate these with the heat of their bodies, the embryonic development of mammals takes place inside the mother's womb as in an incubator. Because the mother suckles her babies and gives them warmth and protection, the initial physical tie between the mammalian mother and her offspring is even stronger than that between birds and their young. This explains why nest-building is neither as widespread nor as important with mammals as with birds.

Nevertheless, there are quite a number of builders among the smaller mammals. Some dig burrows into the ground for shelter, and others excavate extensive subterranean galleries in which they search for food, as moles do. Others, such as the European harvest mouse and the common dormouse, build beautifully worked little nests of blades of grass in fields or thickets which can stand comparison with many a well-built bird's nest. But on the whole, mammals do not handle their nesting and bedding materials with particular skill. In common with certain birds, some mammalian species have discovered that cavities in rocks and trees are suitable places to live in. The most outstanding builders among the mammals are the beavers. Not only do they build fairly elaborate

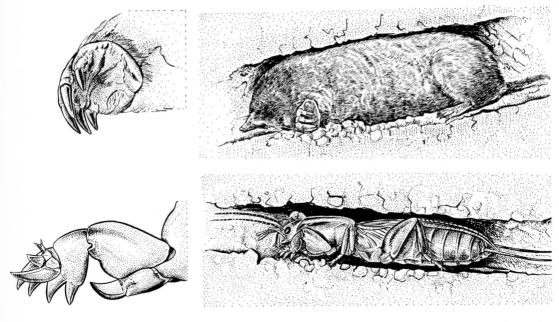

Fig. 93. Top left, drawing of mole's paw; top right, mole tunneling. Below left, mole cricket's claw; right, cricket digging. The shape of the mammal's foot is similar to that of the insect; both are well adapted to their function.

dwellings, they construct large dams which regulate the water level in the streams and lakes where they live to conform to their requirements.

The mole and its subterranean environment

The European mole (*Talpa europaea*) prefers loose fertile soil, but its habitat is not restricted to such favorable conditions. Though common all over Europe, including Britain, it is rarely seen as it prefers to stay below ground. Its body is completely adapted to its subterranean existence. Its front legs are large, turned outward, and equipped with strong claws. Surprisingly similar digging tools are found with the mole crickets (fig. 93). As so often happens, similarity of function has led independently to the development of similar forms in animals of entirely different origin. The mole's head sits close on its shoulders and looks as if it were joined to the body without a neck. But in fact, the number of vertebrae in a mole's neck is seven, the typical number for mammals, and the same as in the neck of a giraffe. Only the length of each vertebra differs between the two species.

The hairless snout of a mole looks like that of a miniature pig. When it burrows through the ground, it loosens the soil with its snout and two front legs and then partly presses the loosened soil against the side of the tunnel with a turning movement of its body like that of a drill, and partly pushes the dirt toward the back with its hind legs. From time to time it pushes through to the surface and throws up a molehill. These molehills serve as ventilators for the network of subterranean passages and enable the mole to reach the surface whenever it wants to. Its small rudimentary eyes hidden under fur, its lack of outer ears, its spindle-shaped body, and its soft coat, the hairs of which fold back easily and do not impede its moving back and forth through narrow passages, are all adaptations to life in dark, narrow tunnels.

The mole's living sphere is, however, not so uncomfortable as it might appear. Under each molehill, the tunnel widens to a small cave about twenty centimeters in diameter; this is comfortably lined with grass and dry leaves. These chambers are used by the animals for resting and sleeping (fig. 94). One must not assume that moles move over wide areas; they usually inhabit a strictly limited territory, measuring about thirty by forty meters, its

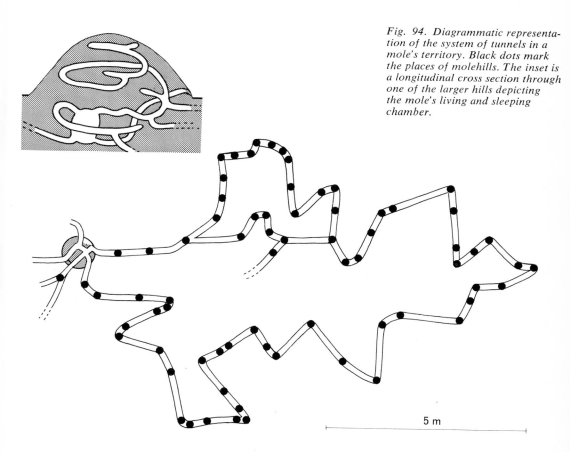

Fig. 94. Diagrammatic representation of the system of tunnels in a mole's territory. Black dots mark the places of molehills. The inset is a longitudinal cross section through one of the larger hills depicting the mole's living and sleeping chamber.

5 m

actual size depending on the nature of the soil and the amount of food available. At a depth of about ten to thirty centimeters, hunting tunnels radiate from the nest chambers in all directions. These also serve for general circulation and may contain one or more additional chambers with nesting material as alternative resting places. New passages are constantly being built, and old ones collapse.

Apart from the mating season, both male and female moles live solitary lives, each in its own domain that is its home for life. However, as their normal life span is only three years, they do not enjoy this possession very long.

The daily course of their activities follows a regular pattern. Actually, one should not speak of a daily course, because day and night matter very little in the dark world of the moles; it does not affect them, when they move about their hunting grounds, whether the sun or the moon is in the sky. Four to five hours of activity are followed by three to four hours of rest or sleep in their soft nests. Their movements have been carefully recorded without disturbing them, by attaching radioactive tracers to their bodies.

Moles require a great deal of food. They eat about the equivalent of their own weight in twenty-four hours. Fortunately for them, a great many creatures that moles like to eat collect naturally in their tunnels. Their chief food is earthworms, but they also consume many insects and insect larvae, millipedes, spiders, and many other animals. More food is obtained by the excavation of new hunting galleries. Occasionally moles leave their burrows and hunt on the surface for snails and insects. Sometimes they will even attack and devour the young of ground-nesting birds. Though moles cannot see, they can find their way about their neighborhood extremely well with the aid of their other senses. Their sense of smell is excellent, their sense of touch is highly developed, and they can hear though they lack outer ears. In times of plenty, earthworms and other food are hoarded in storage chambers or in abandoned tunnels. The moles bite off the front part of the earthworms they wish to store. This part contains the center of coordination for an earthworm's movements and its loss prevents the worms from crawling away. Biting off their heads achieves exactly the same effect with earthworms that the paralyzing sting of digger wasps achieves with caterpillars. The prey is immobilized but remains alive so that the food does not deteriorate. One observer counted 1280 earthworms and 18 white grubs in a single storage chamber. Hoards of this nature are very important for the winter because moles do not hibernate. They continue their regular pattern of rest and activity, except that during periods of severe cold they dig their passages to a greater depth, about sixty centimeters under the soil surface, where frost does not penetrate.

When spring approaches, the female builds a breeding chamber with a particularly soft nest. This is the only time of the year when moles interrupt the solitary tenor of their lives. When rivals meet during mating time, they often engage in violent fights which may end fatally. It is not unusual for the vanquished rival to be eaten by the victor.

The female gives birth to about four young. After about two months, they become independent and leave their mother's abode to find their own territories and dig their own homes. Tunnels made by young animals are not so deep in the ground as those of adults and can be seen as little ridges on the surface. There are several genera of moles in America. Their burrows, their appearance, and their mode of life resemble those of the European mole.

The badger

The mole not only has its home in the depths of the earth, but the subterranean world is also its hunting ground. Few other mammals are soil-dwellers to that extent, but many species excavate their dwellings and their nurseries in the soil. A well-known example is the badger (*Meles meles*), which inhabits a wide area stretching from Britain via Europe and Asia to Japan. It belongs to the same family as the martens (Mustelidae), but is by no means a bloodthirsty predator like the other members of the family. The badger is omnivorous, consuming much vegetable matter, but it also kills many mice and voles and in general refuses little that is edible.

Our country home on Lake St. Wolfgang in the Austrian Alps is surrounded by meadows that are bounded by woods about a hundred meters distant. I remember having seen these meadows thoroughly churned up, as if they had been visited by wild boars. This was the work of a badger which had come down during the night and which had used its pointed snout to compete with the subterranean moles. Badgers leave their burrows, or "sets," only in the evening twilight or at night (pl. 110, p. 272).

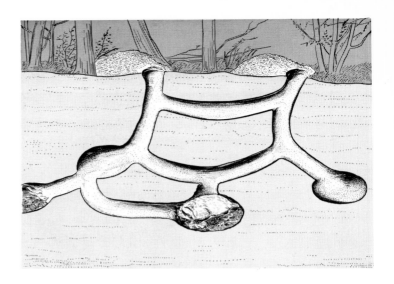

Fig. 95. Cross section through part of a badger's burrow.

A badger's subterranean burrow is usually found within a wood or near its border, in dense scrub or open tree stands. It covers an area of about ten to thirty meters in diameter and reaches a depth of about five meters (fig. 95). It may have three stories of chambers which are connected by a labyrinth of passages; usually it has several exits. On dry nights, the badger will collect whatever suitable bedding material it can find in the neighborhood, such as moss, ferns, or leaves. When the animal has got a pile together it pushes it with snout and front legs into the burrow where the material is used to line the sleeping quarters and, at the appropriate time, the nursery. From time to time, the badger brings its bedding out into the open and, after a few hours, takes it back inside, just as we sometimes air our pillows and mattresses.

Apart from the mating time, a badger very often lives alone in its burrow to which it becomes very much attached. It may gradually extend it, and there are burrows with tunnels of up to a hundred meters in length and with forty to fifty exits, not all of which will be in use at the same time. Such large burrows are, however, usually inhabited by several families of badgers. Foxes, which live in similar subterranean dwellings, often make use of the badgers' enthusiasm for digging and occupy a part of their extensive systems of tunnels and chambers. The badgers do not object, but apparently they do not like the smell of fox and block up the passages between the two parts of the burrow.

Badgers doze much of the winter away in a well-lined, deep chamber, though they do not hibernate in the strict sense of the word. Their body temperature does not fall much below normal as it would in actual hibernation, when the metabolic rate is greatly reduced. During this period, they live on the reserves of fat which they have accumulated in the good months. They wake up quite frequently and search for food during weeks of milder weather.

In the United States and in southern Canada, the place of the European badger is taken by the American badger (*Taxidea taxus*), which is slightly smaller. Its building activities are similar, but it prefers a meat diet. It preys in particular on small rodents, such as mice, voles, ground squirrels, and rabbits. It is not averse to eating insects, but vegetable food is of minor importance.

Rodents as builders

On the whole, the architectural achievements of mammals are not very impressive. The order of rodents is, however, an exception in that it contains, comparatively speaking, quite a number of clever builders. This is probably due to the fact that nature has equipped them with suitable tools. There are, first of all, their mobile *hands*. Everybody, no doubt, has watched with pleasure the skillful movements of mice or squirrels holding a seed or fruit in their little front paws, turning them this way and that while nibbling them. Their other major tools are the long and powerful incisors in their upper and lower jaws, with their chisel-like edges. Though the teeth suffer a great deal of wear when used to gnaw wood or other hard substances, they are never worn down, because, in contrast to the teeth of other mammals, rodents' teeth never stop growing. Let us look at the longitudinal section of a young human incisor tooth (fig. 96, top left). The point of the root is wide open. The nourishing bloodstream has easy access to the pulp cavity in the interior of the tooth. Later in life, the point of the root narrows (fig. 96, center), the circulation of blood to the tooth is impeded, and growth ceases. In contrast, the incisors of rodents retain a wide opening at the base throughout their lives (fig. 96, top right) and growth never stops. It is, however, constantly compensated by wear at the top. The bone-type substance of which a tooth is built up (fig. 96, 2) is covered with an outer layer of an extremely hard and resistant material, the tooth enamel (fig. 96, 1). The

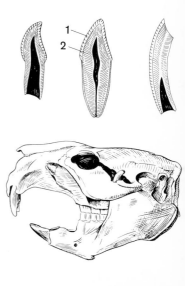

Fig. 96. Incisor tooth: above left, of a young human; center, of an adult human; right, of a rodent. In rodent's teeth, the pulp cavity (black) remains wide open and the nourishing blood stream circulates freely. (1) Enamel of the tooth; (2) bone. Below, skull of a beaver.

Fig. 97. Nest of a harvest mouse in an oat field. An oat stem nearby has been bent and incorporated into the nest for extra support.

incisors of rodents have no enamel on their inner sides. Outer and inner faces therefore wear at an unequal rate and this keeps a knife-sharp edge on the tooth.

The harvest mouse. With front paws and incisors as her tools, the European harvest mouse (*Micromys minutus*) constructs a dainty little nest that can stand comparison with some of the most skillfully woven birds' nests (fig. 97). The builder of this tiny home is one of the smallest rodents in existence, its body measuring only six to seven centimeters in length and its tail almost exactly the same (pls. 101 a and b, pp. 256–57). Found in Europe and Asia, including Japan, its habitats are fields of cereal plants, meadows with long grass, stands of sedge and reeds on the shores of lakes, marshland, or grassy scrub. It lives chiefly on the seed of grasses, cereals, and herbs. But it will also occasionally eat insects, and by no means only the smallest kinds but grasshoppers or cockchafers as well. It climbs about on blades and stalks with great agility and charm, using its prehensile tail as a climbing support. Its nest is usually situated a half meter to one meter above the ground.

The one shown in figure 97 had been built in a field of oats. The mouse had bent one stalk and incorporated it into its nest as a support. While building, it grasps the thin leaves of oats growing nearby or, on other sites, blades of grass or sedge or whatever else is handy, and pulls them through its mouth reducing them to longitudinal shreds with the tips of its incisors. Once a base has been formed by interlacing the leaves of stalks standing together, the rest of the nest is built up in a similar manner and finished off with a dome-shaped roof. If there are no more growing leaves within reach, the mouse bites off leaves in the vicinity, shreds them lengthways, and uses them to strengthen the structure from the inside until it has fashioned a firm dense ball with a lateral entrance hole. Finally, the inside is padded with seed down, panicles of flowers, and shredded leaves. The little builder is amazingly dextrous and can finish a nest in from five to ten hours.

The nest described is a breeding nest made by a pregnant female as a nursery for her young, usually three to five in number. The offspring develop rapidly and can leave the nest after little more than two weeks. The tiny creatures are very lively and often try to push their way through the fabric of the nest. Hence, it is hardly surprising that the nest is in pretty poor shape after housing one litter and that the mother has to build a new nest every time she produces more offspring, which is several times a year.

Sleeping nests are simpler and less dense. They are built also by adult males and by immature animals, though the young ones are not yet very good at it.

Although their small size causes them to lose a great deal of heat in cold weather, harvest mice do not hibernate. Occasionally they build winter nests in soil cavities or other sheltered places and stock them with stores while there is still food about. Sometimes they nest in granaries and barns where they can feed to their hearts' content all

Plate 101 a. European harvest mouse and nest.

Plate 101 b. The nest is no longer as neat as that shown in figure 97 (p. 254). It has already served for raising of a litter and the young mice have pulled it to pieces.

through the winter. In severe winters, many die of hunger and cold.

The dormouse. Life is much easier for the common dormouse (*Muscardinus avellanarius*), whose nest resembles that of the harvest mouse. It spends the cold season fast asleep in its winter nest in some sheltered spot. Dormice, if considered strictly by category, are not really mice at all (family Muridae), but belong to the family of Gliridae, a group that can hibernate for more than half a year. The spherical summer nests of the common dormouse, usually well hidden in undergrowth or in small conifer trees (pl. 102), resemble those of leaf warblers (cf. pl. 84, p. 194). Like them they have a lateral entrance hole.

Plate 102. Summer nest of a common dormouse in a bush. There is a lateral entrance. In structure, the nest resembles those of certain birds. (See nest of leaf warbler, pl. 84, p. 194.)

Wood rats. The American wood rats (genus *Neotoma*) are much bigger than these little creatures. They are rather like ordinary rats in size and appearance and they build much bigger and coarser nests from twigs and brushwood. In fact, their structures can hardly be called nests;

they are small lodges. The dusky-footed wood rat (*Neotoma fuscipes*), a native of the western parts of North America, builds its large untidy twig piles on the ground, usually in thorny thickets, tangles of wild roses, and similar undergrowth. They may be over a meter high and may contain five distinct chambers connected by passages. In addition to a living chamber, a nursery, and storage chambers, there is also a lavatory. Sometimes they build a second home, hardly less elaborate, in the branches of a nearby tree. It probably serves as a refuge when they do not feel safe on the ground.

There are about twenty different species of wood rats, most of them inhabitants of western North America. They are highly adaptable creatures and can live in a great variety of habitats. The white-throated wood rats (*Neotoma albigula*), for instance, like cacti. With great agility, they climb about on these spiny plants, managing not to get hurt, and bite off large chunks that they carry to their homes in their mouths to use as bristling roadblocks on all access routes. They alone can squeeze by unscathed. Working constantly on their dwellings, they keep them in good repair.

Some species have a strong collecting drive. The bushy-tailed wood rats, or pack rats (*Neotoma cinerea*), which resemble squirrels because of their long hairy tails, not only hoard food in autumn, but also all kinds of glittering objects they can get hold of. Tin cans, bits of glass, silver spoons, knives, and other things taken from farmhouses have been found littering the ground in front of their lodges. Their activities remind one of bowerbirds, except that these birds clearly use such objects for courtship displays. No one knows why wood rats collect such apparently useless, inedible things.

Squirrels.   The nests of squirrels are similar, though not so well organized inside. Squirrels are probably the best known and best loved rodents. They are quite common in public parks so that even people in big cities may get pleasure from watching their skillful climbing and delightfully trusting behavior.

Some of the (European) red squirrels (*Sciurus vulgaris*) are red and some are black. Their color may vary between regions, but in many areas both types are found. Sometimes both colors appear in one litter. Their natural habitat is the forest. Their sharp claws enable them to run up the trunks of trees and to move nimbly about the branches. Occasionally they make long jumps from one tree to another using their bushy tails as rudders. Their main sources of food, namely, nuts and the seed of conifers, are also plentiful in forests and bush. To get at conifer seeds, they pick the cones and pull off the scales. Great quantities of scales and of the denuded cores of cones can often be found lying under conifer trees visited by squirrels. Berries and other fruits and also insects form part of their diet. Unfortunately, they also like eggs and baby birds and often rob birds' nests. For their own nests they choose the crowns of trees, preferably some fork near the trunk where the nest will not be too severely shaken in a strong wind (fig. 98 and pl. 111, p. 272). Occasionally they save themselves the trouble of collecting twigs for the nest base and choose a bird's nest of suitable size,

Fig. 98. Main nest (dray) of a squirrel.

such as a crow's nest or the nest of a nuthatch. First, they repair it and stop up holes where necessary. Then they raise the wall and build a roof, ending up with a spherical nest, or "dray," about fifty centimeters in diameter and thirty centimeters in height. Usually they detach stronger branches needed for the outer structure from their nesting tree by biting or breaking them off; less often they collect them from the vicinity. They insert the new material between the existing branches and move it about with teeth and paws until it is firmly wedged in place. The inside of the dray is lined with grass, bark fiber, and other soft materials, and the floor is covered with a finely shredded cushion of grass, moss, lichens, and the like. These furnishing materials are often collected from greater distances and carried in the mouth in the form of bundles or balls. A small entrance hole is left on one side. Sometimes the completed nest is disguised on the outside by twigs still bearing leaves or needles. The building of the dray takes from two to five days. Where old and decayed trees can still be found, squirrels sometimes nest in holes and save themselves the trouble of building a strong outer structure.

Smaller and less carefully constructed nests that serve as places of rest and refuge are often built in other parts of their territory.

In the mating season, males and females jointly use the main nest that serves for sleeping and for the raising of the litter. But even before she gives birth, the female chases the male away from the nest. She alone looks after her offspring. She keeps them clean, turns them, and licks them all over. She also removes all excreta from the nest. After three weeks, the family occasionally ventures out onto a nearby branch. At that time the mother still feeds her young and protects them, but soon after, they are weaned. After eight to ten months they are sexually mature, having become independent much earlier.

Squirrels do not hibernate. It is true that they doze in their nests for several days if the weather is rough or very cold, but they wake up whenever it is milder, and then they need food. Their nests do not contain any storage chambers, but they do build up hidden hoards in autumn when food is plentiful. A preferred place for the hoarding of nuts and hazelnuts is a hole dug at the foot of a tree. They push a nut into the hole with their snouts, hammer it down with their upper incisors, and cover it up with soil and leaves scraped together and pressed down with their front paws. Many a nut disappears into the soil in this fashion, and many such caches are established in the ground and in other hiding places. Though this looks like intelligence and foresight, the squirrel does not consciously make up its mind to store food in order not to starve in winter. It acts in this way whether or not it has ever experienced a winter, obviously compelled by an innate drive. That this is so can be observed in tame squirrels. They will fetch a nut from the fruit bowl and search round the room. In one corner after another they will go through the motions of digging, of pushing the nut into a hole, hammering it down, and covering it with non-existent material, all in the right sequence, and then start again with the next nut, as if they had been completely successful. The fact that their innate pattern of activity cannot achieve the desired result in such an unnatural environment does not deter them at all.

Many of these caches are never found again after months have passed and snow has covered them up. However, they are not quite useless because new nut trees and hazel shrubs grow from the forgotten hoards, so even if the hoarding squirrels themselves do not benefit, future generations of squirrels will.

The same genus is represented in eastern North America by another species, the gray squirrel (*Sciurus carolinensis*). These are a little larger than the red squirrels but resemble them so closely in their behavior that a European visitor quickly feels at home in American parks. They have also become established in Britain and have almost completely ousted the native red species. The American red squirrel (*Tamiasciurus hudsonicus*), which is smaller, is found from the east to the west coast of North America. In high northern latitudes, it hibernates. Other genera and species of squirrels live everywhere in the world except the Australasian region.

Marmots. Some rodents excavate passages and chambers in the soil like the badgers. This is the way of the marmots, whose thirteen species and subspecies are distributed over wide parts of the northern hemisphere. Strong claws on their front and hind legs equip them superlatively for digging.

An alpine marmot (*Marmota marmota*) is about the size of a brown hare, but it is more thickset and has shorter legs. In the Alps, it inhabits regions between one to three thousand meters. It excavates its burrows, which usually reach a depth of three meters, on sunny alpine meadows or not too forbidding screes. Marmots are social creatures and live in colonies comprising sometimes only a handful of individuals but sometimes fifty or more.

Mountain farmers, though usually selfish enough in their endeavor to protect their lands from possible damage, tolerate the marmots and do not begrudge them the grasses, herbs, and roots which they eat on these upper pastures. Thus observant walkers can enjoy watching their delightful behavior. Around the turn of the century, these animals had been nearly exterminated in large parts of the Alps, chiefly because folk medicine attributed very special healing powers to their fat. In recent years, they have been successfully reintroduced in many places, have multiplied greatly, and have spread.

Part of a large burrow of the kind that serves as winter quarters for a family of parents and their young is shown in figure 99, bottom. When the animals dig the narrow tunnels, which are just wide enough for them to get through, they push the excavated soil under their bodies with jerking movements of their front legs, and then throw it backward with their hind legs. The tunnel system branches extensively and its many exits frequently lie under stones or boulders. One passage, which may be quite long, leads from the main entrance to a large central chamber which is softly padded with hay (1 in fig. 99). Smaller chambers may occur along other passages. Short dead-end passages (2) serve for their excreta.

At the approach of winter, the marmots collect dry grass and herbs and carry large bundles of these to their homes in their mouths (pl. 112 a, p. 272). As much as twelve to fifteen kilograms of hay have been found in one burrow. When they have finished these preparations, they pack large stretches of their tunnels, sometimes several meters, with hay, soil, and stones. Such a long and solid blockage provides excellent protection against the cold. The whole family huddles together, each animal rolled into a little ball, and hibernates in the main chamber. During hibernation, which may last half a year or more, the marmots' metabolic rate is much reduced, and their body temperature falls to 5°–7° C. (41°–44.6°F.) Every three or four weeks or so, they wake up for a short while to urinate. Their metabolism revives slightly, leading to a temporary rise in body temperature and, thereby,

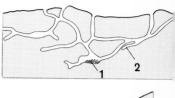

Fig. 99. Marmots' burrow. Top, a summer burrow; below, winter, or permanent, burrow. (1) Main chamber; (2) a short passage for feces.

to a slight warming of their chamber, but they soon go to sleep again.

When spring at last comes to the mountains, the doors are unblocked, the burrows are cleaned, and all dirt is thrown out. Suddenly the whole picture is changed. The lethargy of hibernation is replaced by the lively business of mating and the procreation of new life. Mountain summers are short, and there is no time to waste if the young are to be strong enough by the onset of winter to withstand its rigors. Before her confinement, the female marmot blocks the entrance to her living chamber with hay, and thus sees to it that she is not disturbed. The young are weaned about six weeks after birth and around that time they leave the burrow for the first time. They are watched over by their mother for the rest of the summer, and they spend their first winter hibernating in their parents' nest. The following summer, they gradually gain their independence and excavate their own burrows.

With the onset of warmer weather, marmots often move to existing summer quarters at higher altitudes or build themselves new summer burrows (fig. 99, top). These are not so deep as the winter burrows and tend to have more entrance holes. But the habit of summer migration is not universal. In some areas, the marmots stay in the same burrows all the year round, and it would, therefore, be more correct to call these permanent burrows and not winter burrows.

Marmots leave their burrows only in daytime (pl. 112 b, p. 272). Observant walkers in the mountains may have seen older marmots standing on some stone or

Fig. 100. Marmots. Sentinel at the entrance; in the background, marmots at play.

boulder close to an exit hole acting as lookouts, their sturdy upright figures supported by their strong, muscular tails (fig. 100). Usually, however, the only sign they will have of the marmots' presence will be the shrill whistle which has sent all the animals within earshot scuttling into their burrows before they could be seen. Actually, one should not talk of a marmot's whistle. Their warning signal is a sound produced in the throat.

The sentinels keep a sharp lookout not only for people, but also, and primarily, for their archenemy, the golden eagle. If there were no lookouts, marmots would be easy prey because they love to play and become so absorbed in their games that they pay no attention to danger. The young romp for hours in the open air, and the older animals often play with them or even with each other. Sometimes they will roll down a slope just as we did as children, either alone or two together in a tight embrace. Sometimes many join together and play group games. This carefree life would hardly be possible without the protection of a watcher whose shrill signal tells them to disappear.

A warning might easily come too late if special provisions for such rapid disappearance did not exist. In addition to their ordinary entrances, the larger burrows usually possess vertical tubes into which a fleeing animal can dive and so get to the bottom of the burrow the quickest way possible, that is, by falling. Also, whenever marmots move more than twenty meters away from their burrow in search of food, they dig themselves emergency escape tunnels. These are short passages, often hardly a meter in length, widened sufficiently at the end to enable the animal to turn around. With their summer and winter residences, their vertical diving tubes and outlying flight tunnels, the burrows of marmots are somewhat more complex than the usual burrows of mammals.

There are several species of marmots in North America. The hoary marmot (*Marmota caligata*), which has a very light fur, is found chiefly in the north and northwest. The yellow-bellied marmot (*M. flaviventer*) occurs in the west, in an area reaching from northern New Mexico to British Columbia. Both are smaller than the alpine marmot, but their habits and burrows are similar. The smallest and most widespread species is the woodchuck (*Marmota monax*). In contrast to the alpine marmot, this little animal leads a solitary life for most of the year. Its burrows are usually found in open wood-

Plate 103. A beaver dam. Behind it, the dammed-up pond and a lodge. In the background, left, another dam. Rocky Mountains. (See p. 273.)

land and at forest borders. This creature may not be popular with farmers because of the considerable damage it does to gardens and fields, but motorists young and old delight to glimpse the furry woodchucks grazing on grassy slopes alongside scenic highways.

Rodents are clever and highly adaptable builders. Their styles include not only the delicately woven nests of harvest mice, the coarser twig nests of wood rats, the drays and hole-dwellings of squirrels, and the subterranean burrows of marmots, hamsters, and many other species, but also the entirely different structures which the beavers create in watery surroundings.

Beavers. The beaver (*Castor fiber*) is one of the largest and heaviest rodents in existence. It may weigh up to thirty kilograms. It is an excellent swimmer, fast and skillful under water, and if frightened can stay submerged for up to fifteen minutes before it cautiously comes up for air. Its webbed hind feet and broad tail, which it uses

266

as a rudder, are exceptionally useful in the water; the beaver is much clumsier on land. Because of its swimming skills and because of the scales that cover its tail, the beaver was regarded, for dietary purposes, as "fish" by the Roman Catholic Church, which permitted consumption of its palatable flesh during Lent.

Large beaver colonies used to be widespread along the banks of rivers, streams, and lakes in the northern forest belt from the United States and Canada to Europe and Asia. Excessive persecution has decimated their numbers. In Britain they became extinct in the thirteenth century, and in the nineteenth century only small residual populations existed in continental Europe, such as those on the Rhone and Elbe rivers. Their numbers have also declined rapidly in North America and Russia. They were hunted and trapped chiefly for their fur which was much sought after, and also for the large glandular pouches near their genitals which contain castor (castoreum), a substance used by the beavers for marking their territory.

Plate 104. Beaver dam shown from the dammed-up side. A lodge in the foreground. The two photographs give an idea of the solidity of the structure. Cotton Wood Creek, Colorado.

In medieval medicine this was considered to be a cure for virtually every ailment under the sun, and large sums were paid for it. Much serious damage to beaver populations has also been done by extensive river management schemes. These have been carried out with complete disregard for the habitat requirements of those very animals that previously had altered the beds of rivers to suit their own needs. For it is well known that beavers are experts not only in the building of dwellings but also in hydro-engineering, and have performed tremendous feats in this line long before man attempted anything of the kind. Beavers are now actively protected in most countries, and this has led to a welcome increase in their numbers. They have also been successfully reintroduced into areas from which they had long ago disappeared.

Beavers vary their dwellings in accordance with local conditions. All they do on large rivers where the water level does not fluctuate appreciably is to dig an upward-sloping tunnel into the riverbank which they widen into a subterranean chamber about one meter in diameter and half a meter high (fig. 101, top). Such a chamber has at least two, and sometimes more, entrance tunnels—something that cannot be seen in the diagrammatic drawings. The entrances are all under water, and a feeding chamber is close to the entrance. Beavers like to feed near the water's edge at night, either in the open or on the small artificial beaches of their tunnels. They are very clean animals, and dispose of all leftovers straight into the water.

Should the level of the water rise, soil is scraped or gnawed off the ceiling of the chamber and used to build up the floor. If the roof layer becomes too thin, soil and twigs are heaped on top. Should the waters rise even further, the animals add to this "lodge," as it is called, and move their living chamber into it (fig. 101, center and bottom).

In shallow, slow-moving waters, such as in a dammed-up pond of their own making, they build the lodge so that it is surrounded by water. By piling up on the pond bottom branches, twigs, and pieces of bark, which are all so saturated with water that they do not float away, they build an artificial island. This they enlarge by adding more sticks and twigs until a height of two and sometimes three meters above the water's surface is reached. Inside this pile, they hollow out their living chamber and passages, all of which end below the water's surface (see pls.

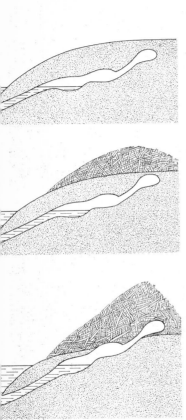

Fig. 101. Beavers' dwelling on a bank. Top, toward the front of the entrance tunnel, the feeding chamber; higher up, the main chamber. Center: if the level of the water rises, the chamber is moved upward and a pile of twigs and soil is heaped on top to protect the roof which has become very thin. Bottom: should the water rise even higher, the pile on the surface is enlarged and the main chamber is moved into the "lodge."

Plate 105. Beaver swimming to building site with tree branch. (See p. 276.)

270

Plate 106 (left, top). Beaver dam. Rocky Mountains. (See pp. 272, 273.)

Plate 107 (left, bottom). Beaver lodge. Rocky Mountains. (See p. 273.)

Plate 108. Beaver felling a tree. (See p. 275.)

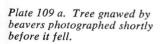

Plate 109 a. Tree gnawed by beavers photographed shortly before it fell.

Plate 109 b. Stump of tree gnawed by beavers, shown after tree had fallen. The traces of the animals' incisors are clearly visible between the bark and splintered center. (See p. 276.)

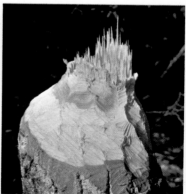

Plate 110. *A badger leaving its burrow. (See p. 251.)*

Plate 111. *A squirrel's nest. (See p. 260.)*

Plate 112 a. *Marmot carrying hay to the burrow. (See p. 263.)*

Plate 112 b. *Marmot at entrance of burrow. (See p. 264.)*

on pp. 266, 267, 270). The walls are carefully caulked with mud and clay, but not entirely. Part of the domed superstructure is left loose to allow sufficient ventilation. On cold days little clouds of steam rising from the top of beaver lodges are indicators of life and warmth inside.

Many animals can construct homes from branches and twigs as beavers do. But the use of these materials to build across a stream or river a dam that transforms a shallow stretch of water on the upstream side into a calm lake of sufficient depth for their lodges and for their swimming and diving activities is an achievement not matched by any other animal. They set to work on a locale which in its natural state is not fitted to their needs and thereby transform the character of the upstream water and of the whole landscape in such a way that it becomes suitable for their settlements.

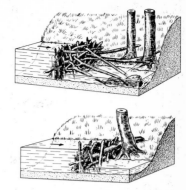

Fig. 102. Examples of construction techniques used by beavers. Diagrammatic representation.

They go about their tasks like experienced craftsmen. At the site of a future dam they ram strong sticks into the bottom of the stream or river bed with considerable force, push twigs in the interstices, weighing them down with heavier sticks and supporting them against the current with forked branches; or they provide the necessary resistance by the insertion of crosspieces. The structure is anchored to growing trees or boulders, and reinforced with heavy stones brought to the site (fig. 102). Gaps are filled in with fine twigs and reeds or other small material and covered with mud or clay so as to make the finished dam completely watertight. The underwater part of the upstream wall is steep and smooth. In front of it, there is a large, deep pit from which the beavers have taken their building material. This pit has, moreover, the effect of reducing the speed of the current and thus protects their structure. The downstream wall is a tangle of coarse branches anchored at the bottom and to the sides of the stream. The crown of the dam is slightly lower at the sides than at the center so that the dammed-up water runs over it near the banks. The details of the design are flexible, as is the size. Dams may be built across narrow streams and brooks but also in large rivers (pl. 106, p. 270.). The largest recorded dam in the Voronesh region of Russia is 120 meters long, 1 meter high and 60–100 centimeters wide. In the swamps of the Mississippi basin, beavers build dams several hundred meters in length. It is the only way that they can produce lakes of sufficient depth for their needs. Possibly the beaver dam on the Jefferson River in Montana (United

Plate 113. Beaver repairing a breach in the dam. Kanawahka Lake, New York.

States) is the largest ever built. It is seven hundred meters long and can bear not only the weight of a man but also that of a rider on horseback.

Large dams are maintained by generations of beaver families over a period of many years, perhaps even centuries, and are constantly adapted to changes in the water level. If the water rises so high that it threatens to submerge the living chambers of their lodges, the beavers do not attempt to alter conditions at the place of flooding where repairs could not effect long-term relief. They tackle the situation at the dam, which may be a considerable distance away, by lowering the part of the crown that the water runs over, which causes the level of the lake to fall. Should the water level fall too low, they raise the height of the dam and improve its density where necessary. If the drop in the water level has been brought about by a breach in the dam, they quickly find and repair the damage (pl. 113). A skilled hydro-engineer could not regulate the water level any better.

274

The construction of dams and their constant maintenance represents teamwork on a large scale. Not only the members of each family but the families of a colony work together effectively. Because they need a great deal of wood for their dams and lodges, the beavers tend to colonize rivers or streams that run through wooded bottom lands with plenty of undergrowth, and to avoid water courses whose banks are treeless. Aspens, willows, and poplars (cottonwood) are the preferred types of lumber, but these remarkably efficient woodcutters are not deterred by the hardwood of oaks. The edges of their chisel-shaped incisors are so sharp that beaver teeth were used by some American Indian tribes as their most valuable cutting tools.

It takes a beaver only a few minutes to bite through a thin stem. Holding its head sideways, it makes an oblique cut. Stems with a diameter of twenty centimeters or more are gnawed at from all sides (pl. 108, p. 271), often by two animals working in turns. This methodical attack

Plate 114. A beaver transports branch from felled tree along the beaver trail toward water. (See p. 276.)

275

produces a regular, all-round indentation, shaped like an hourglass, which eventually causes the tree to fall. The animals' teeth marks can be clearly seen on the stumps (pls. 109 a and b, p. 271). It may take several nights to fell a tree that is large and hard. Beavers have good hearing and can tell by the cracking noises of the wood when a stem is about to break. Then they suddenly rush away and dive into the water to avoid being hit by the falling tree. Occasionally, though rarely, they do not get away in time.

Suitable felling tools alone are not enough for good forestry. Extraction lanes are needed to transport the lumber to the places where it is needed. Trees standing on the banks present no problem in this respect. They either lean toward the light or have an excentric crown that is heavier on the side exposed to light, and they will automatically fall into the water when they are felled. The beavers cut them into branch and stem sections and float them to their building site with ease. Trees standing farther away from the bank are reached by well-trodden beaver paths. The animals can carry smaller stems in their mouths to the water's edge and swim with them the rest of the way (pls. 114, p. 275, and 105, p. 269). Dragging larger pieces of wood to the water is certainly hard work, but the beavers carefully remove all obstacles from their extraction lanes and smooth their surface. They have also devised another way to solve the transportation problem. Where the ground is level, they dig special canals to the lake; these may be up to fifty centimeters deep, sufficient for the beavers to float the wood in the water while they swim.

The beavers' teeth are their main tools for felling trees and cutting them into pieces. Their *hands* are indispensable for building. The hand-like forepaw of beavers has only a poorly developed thumb, but the little finger takes over thumb functions, enabling the animals to get a firm grip on sticks and branches (fig. 103). With these hands, they pick up smaller sticks and brushwood to be held in their mouths until they have collected a bundle; then they swim back to the building site. The stones needed for strengthening their dams they pick up and lift with both hands. They also use their hands as shovels to take up small brushwood and loose bark debris from the bottom of the river. Holding the material between arms and chin, they swim underwater to the building site. Then they carry the load to the top of the lodge or the crown of the

Fig. 103. A beaver's hand and a human hand. The beaver's thumb is poorly developed; the little finger has taken over the role of our thumb.

276

dam by walking upright on their hind legs. The clay or mud used to make their structures watertight is firmly pressed into place with these skillful hands.

Beavers are not only remarkably clever in handling wood and using it for building. Their ecological association with woody vegetation goes further. They are herbivorous and are fond of the thin green bark and foliage of freshly felled trees. In fact, bark, softwood, and leaves form the main part of their diet, to which they add various herbs and aquatic plants in the warmer months of the year. However, neither their feeding habits nor their building activities are easy to observe because they rarely leave their dwellings before dusk.

Beavers do not hibernate. Yet they may be cut off from dry land during periods of hard frost when the water is covered with a solid layer of ice, for one must remember that the exits of their burrows are all below the water's surface. Such an emergency does not find them unprepared. In the fall, they cut down trees to be used not for building but for winter stores. They collect large quantities of branches and brushwood which they anchor in heaps at the bottom of the water in the vicinity of their lodges, sometimes next to their front doors. In places where the water is dammed up, they make holes in the dam as soon as the pond is completely frozen over in order to lower the water table. This creates an air space under the ice, which enables them to breathe and swim on the unfrozen surface, and thereby solves the problem of the transport of provisions stored at some distance from their lodges. There is no need for the inhabitants to starve.

Perhaps I ought to tell who lives in the lodge. In the first place, there is an adult couple joined, so it seems, in a lifelong union. In winter, their offspring of the last year and the year before usually live there, too. This creates no problems of overcrowding because beavers, whose cleanliness has been mentioned before, never foul their homes and usually defecate in the water. In spring, when the female beaver is about to give birth to a new litter on a bed of wood shavings, the rest of the family have to move out for a time. The two-year-olds now start out on a life of their own and make room for the new offspring. Beavers may reach an age of ten to fifteen years.

In comparison with other rodents, beavers are endowed with an exceptionally well-developed brain. Considering their many and varied activities, this is hardly surprising. Nonetheless, the principles of their art are theirs by inheritance. Young beavers, reared in captivity, which have never seen a beaver pond, have been observed felling trees in their compound as if they had been taught their job by experienced beaver craftsmen. They also erected a typical lodge from branches and stones and whatever finer materials they could find, and built a perfect, watertight dam in moving water, without mistakes or false starts. This does not mean that, living with their families in their natural habitat, they do not acquire additional skill, profiting from example or experience. There is still more to be discovered by observant biologists.

Apes

Of all living animals the chimpanzees are closest to man. This holds true whether one bases one's assessment on their anatomy, their physiology, or their mental capacity. A visitor to the zoo feels strangely fascinated by them. He discovers traits akin to his own in each facial grimace, in each expressive movement, and in their entire behavior. However, most zoo animals were born in captivity and grew up with humans. May they not have learned to "ape" us?

In recent years chimpanzees and other apes have been studied very thoroughly in the field. In some respects the wild animals appear even more human than those in the zoo. Some fascinating new insights have been gained into their habits and family life, but no remarkable building activities have come to light. Their building consists merely in the construction of simple nests for sleeping. We owe much detailed information on their building habits and other behavior to a young Englishwoman, Jane Goodall (now Mrs. van Lawick-Goodall), whose findings have been published in *In the Shadow of Man* (Boston and London, 1971). For ten years she studied the apes in the Gombe Stream Chimpanzee Reserve (now a national park) close to Lake Tanganyika in central Africa, where these animals are protected and allowed to live in peace. In the beginning, her patience was put to a severe test. The apes were shy and refused to let her get a glimpse of their lives through the curtain of forest trees. But gradually they got used to her unobtrusive presence and tolerated her near them without letting her proximity affect their behavior. Admittedly, it took her over a year to reach this stage.

Except for infants, each individual chimpanzee builds

a fresh sleeping nest every night, usually not before dusk. Normally this work takes no more than three to five minutes. The ape chooses a firm foundation in the crown of some tree, such as an upright fork or crotch or two horizontal branches. Then he reaches out and bends over smaller branches onto this foundation, keeping each one in place with his feet. Finally he tucks in the little leafy twigs growing round the rim of his nest and lies down, pressing the whole cushion down with his body (fig. 104). Quite often a chimpanzee sits up again after a few minutes and picks a handful of leafy twigs which he puts under his head or some other part of his body before settling down for the night. Miss Goodall climbed up into some of the nests after the chimpanzees had left them. In some of the nests she found evidence of quite compli-

Fig. 104. A chimpanzee lying in its sleeping nest.

cated interweaving of branches notwithstanding the speed with which they were constructed. The nests of healthy animals were never fouled with dung. Later, when the chimpanzees allowed her familiar figure to come near their nests, she observed that they were always careful to defecate and urinate over the edge of their nests, even in the middle of the night.

After a female has given birth, she builds a particularly large sleeping nest. This needs somewhat more time, partly because the mother has only her feet and one free hand to work with, as the other hand holds the baby. In one case under observation the mother took eight minutes instead of the usual three to five. An infant stays with his mother for a long time. It is breast-fed by her and sleeps in her nest. This stage normally lasts about four years, but a close tie with the mother remains for several more years. The father does not take part in family life at all.

The urge to build a nest develops long before the infant leaves its mother's sleeping nest. Jane Goodall noticed the first attempt in a little male chimpanzee aged ten months. He bent a little twig over and sat on it on the ground in the approved manner. Then he bent a handful of grass stems onto his lap. During the next few months he became more and more proficient and, like other year-old infants, he often made a nest when he was playing by himself in a tree. Sometimes he lay in it for a short while, but more often he just bounced around in it, often destroying it, and then, after a few minutes, making another. Thus, when a youngster is four or five years old and ready to sleep on his own, he is already skilled in nest-making techniques through constant practice.

Observations in other areas indicate that not all chimpanzees build a new nest each night. When they do not roam about, many animals use the same tree several times and often the same nest as well.

The sleeping nests of orangutans and gorillas resemble those of the chimpanzees, except that gorillas live most of the time on the floor of the forest and often build their nests on the ground. They make a new nest each night, usually in a new place.

There is no doubt that observation in the wild is the best way to learn about the natural behavior of these animals. But as far as their building skills are concerned, important and highly interesting information about apes has been gathered from chimpanzees in captivity.

At the time of the First World War, Wolfgang Köhler

carried out experiments with chimpanzees at the Anthropoid Station on Tenerife (Canary Islands). These experiments have become widely known through his book *The Mentality of Apes* (translated from the second revised edition, New York and London, 1925). They prove conclusively that chimpanzees can solve certain problems by means of intelligent reasoning. This distinguishes their mental processes from those of other animals whose actions are triggered chiefly by instinct. Köhler's results have since been confirmed by many other investigators. Here I want to discuss only a few of his experiments which have a bearing on our theme.

In one instance, a banana was suspended so high in the cage that the animals could not reach it by jumping. An open wooden packing case was placed not far from the spot. At first, the chimpanzees took no notice of the case. Suddenly, one of them stood still in front of the box, seized it, tipped it hastily straight toward his objective, climbed on to it, and with a jump got his banana. In other experiments, the same animal piled two boxes on top of each other to reach his objective.

Another time, the desired object was placed even farther out of reach in a room containing several boxes. A male chimpanzee named Sultan placed a heavy box flat under the target and up-ended another box on top, but even that did not do the trick. Standing on his tower, he searched around until he caught sight of yet another small box. Carefully he climbed down, picked up the box, climbed back with it, and added it to the tower.

These experiments revealed great individual differences between chimpanzees. Some were gifted, others less so. One chimpanzee, a young and vigorous female, was particularly good at solving this kind of problem. She did not lose patience as quickly as many of the others. In one of their large open-air cages a tempting morsel was hung exceptionally high. This animal, which had previously built a tower of three boxes, now managed to add a fourth when she discovered that this was the only way for her to reach the fruit (fig. 105).

The chimpanzees' search for suitable objects to build with, the way they brought them to the right spot and piled them up, was fascinating to watch and conveyed the impression of deliberate and purposeful action. But the animals did not seem to have any sense of statics, no conception at all of balance. They placed their boxes on top of each other in a completely haphazard manner;

Fig. 105. A chimpanzee builds a tower from packing cases in order to reach fruit hung beyond his reach.

quite often the one on top protruded so much over the one below than any human observer would have worried at once about the stability of the swaying tower. From time to time a tower would collapse, bringing its builder down with it before he had reached his goal. "Thinking" is not always the best way to success after all. It is so easy to overlook one link in a causal chain. Instinct, acquired and consolidated over hundreds of thousands of years, is a more reliable guide to the right action.

How to build a sleeping nest is something apes know by instinct, and these instincts express themselves in playful nest-building by chimpanzee infants long before they need a nest of their own. In the Munich Zoo an adult orangutan was observed similarly building a nest, though he had no material whatsoever to do it with. He sat on a raised board in a corner of his cage, picking up non-existent "twigs," bending them toward him, and carefully pressing them down with the back of his hand. The way the adult ape performed this sequence of movements, as it were *in vacuo,* is evidence of the instinctual nature of his behavior. It is a close parallel to the behavior of our tame squirrel which when autumn came would carry a nut from the fruit bowl to the floor of the sitting room and there would go through all the motions, in the right order, which would have been called for to cache a nut in the ground.

The structures that the chimpanzees created from their packing cases may have been far from perfect, but they demonstrated a spontaneous solution of a novel problem carried out with unnatural building materials. No instinct guided the chimpanzees' behavior. Hence not all the animals were equally able in solving the problem, and some never managed it at all. But the shaky towers of those who succeeded represent original mental achievement, and as such stand on a much higher level than the perfect masterpieces of, say, the spider's web, woven by its creator under the unerring guidance of her innate skill.

Conclusion

Our brief architectural survey has taken us through the whole of the animal kingdom, from the most primitive to the most highly developed species. The word "architecture" has been used in its widest sense to include the skeletal formations built up inside an animal's body without the active use of tools, such as an animal's mouth or legs. In this realm we can hardly speak of progressive development. The structures formed in microscopic protozoa admirably fulfill their vital functions of protection and support and, at the same time, exhibit an exquisite beauty (fig. 5, p. 8). The skeletons built by the coral polyps, which are among the lowest of multicellular animal forms, are no less beautifully shaped (pls. 7 and 11, pp. 18 and 19), and the coral reefs erected by these small organisms in thousands of years of unconscious building are mighty monuments of a size that dwarfs the pyramids of the Egyptian kings. The protective and supporting skeletons of primitive siliceous sponges (pl. 3, p. 17) are collected by divers from the depths of the ocean because their beautiful shapes are enjoyed as works of art. Scientists take it for granted that these structures perform a vital function in the lives of their owners—only what is of proven biological value will develop and survive over long periods. The fact that they appear at the same time as objects of perfect beauty is something I gratefully accept as a gift of nature without wishing to philosophize about it.

The position is different when we consider the architectural achievements of animals that are actively engaged in building. Worms do not build arbors like bowerbirds. Here a higher level of building behavior is usually associated with a higher level of biological organization. In general, this is brought about by the development of astounding instincts. Spiders build their webs (fig. 14, p. 28, and pl. 14, p. 37), bees their combs (pl. 48, p. 80), and termites their mounds (pls. 58 a and b, pp. 130–31; pl. 60, p. 134; pl. 61, p. 135) without being shown how, and without the need to profit from experience. The art of building, which their species have acquired through long periods of phylogenetic development, is passed on to the builders of today in their genetic make-up.

Among vertebrates, things have taken a slightly different course, although here, too, building activities are ruled by inherited instincts to a greater extent than many people believe. However, among certain highly developed birds and mammals, the factor of individual achievement is added, and an animal's own experience may lead to exceptional individual soluticns.

The purposes of the animals' building activities vary. Some build traps for catching food: the primitive pits of the ant lion and the highly sophisticated webs of the garden spider exemplified this. But most animals build to provide protection for themselves or their offspring.

The most primitive form of shelter is a cavity, such as a hole in the ground or the inside of a hollow tree. Some builders make their own cavities, or adapt existing ones to suit their requirements. Indeed, this is probably the most widespread building activity. Many digger wasps lay their eggs in carefully excavated burrows which they stock with food for their larvae and then skillfully block up (fig. 24 b, p. 53, and pl. 30, p. 58). Solitary bees often build breeding chambers in the soil. Ants' nests (fig. 45, p. 106), and termitaries (fig. 58, p. 138) are usually started below the surface. The kingfisher and some parrots nest in holes in the ground. Badgers (pl. 110, p. 272) and marmots (fig. 99, p. 263) also live in a subterranean domain. These are only a few examples. The first human beings, too, lived in caves before they started to build solid huts and thereby created the first real dwelling-houses.

Long before that happened, however, certain animals had been able to build themselves dwellings to live in, as distinct from simple holes scratched out of the ground. Their styles were immensely varied, ranging from the mobile homes of caddis fly larvae and bagworms (fig. 21, p. 48, and pl. 20, p. 39) to the skillfully wrought nests of the weaverbirds (pl. 91, p. 214; and figs. 79, p. 208, and 80, p. 209) and the decorated arbors of the bowerbirds designed as places of tryst (pls. 96 to 99, pp. 234 to 236). Selected examples of all these fill the pages of this book.

The great variety of the materials used by animals to build their dwellings and other protective structures will hardly surprise us. I find it more remarkable that often very different groups have found similar solutions to their building problems. A thick layer of foam to hide oneself or one's offspring is a simple and effective means of protection. Several animals have made this "invention" independently of each other. They belong to entirely unrelated groups and, moreover, use quite different methods to produce their foam. We need only to recall three examples: the froghoppers produce their foam cover, the cuckoo spit, by exhaling air into a drop of viscid fluid which they produce around their bodies (fig. 23, p. 50, and pl. 25, p. 40); labyrinth fishes fashion bubble nests for their broods by snapping up air at the surface of the water and releasing it, mixed with a slimy salivary secretion (fig. 67, p. 158, and pls. 66 and 67, p. 153); the Javanese flying frogs whip up foam with their hind legs (fig. 72, p. 166).

Saliva is used by fishes to prevent the bubbles of their foam nests from disintegrating. A quick-hardening saliva is sometimes used as a building material by itself. This is done by certain termites, for example, and by the swiftlets among the birds. Paper wasps use as their basic raw material small particles of wood gnawed off from trees or fence posts, and mix them with their hardening saliva to produce the paper from which they fashion their nests (pls. 32 to 36, pp. 59 to 60). These wasps are not the only manufacturers of paper. Some termites use their saliva (or their excreta) as a cement to make a cardboard-like substance from wood particles, and some ant species build carton nests (pls. 56 and 57, p. 100) by still another method. The jet ant (*Lasius fuliginosus*) saturates the wood particles with a sugar solution, thereby creating a nutritive substrate for a fungus whose hyphea bind the particles together and strengthen the nest structure.

Many vertebrates fashion their dwellings from vegetable substances, threads, and fibers woven together with varying degrees of skill. Such structures include, at one end of the scale, the primitive nests of sticklebacks (fig. 69, p. 161, and pls. 68 a and b, p. 153) and wrasses (fig. 70, p. 163), and at the other, the beautifully wrought nests of penduline titmice (pl. 87, p. 196), weaverbirds (figs. 79 and 80, pp. 208 and 209, and pl. 91 a, p. 214), and harvest mice (fig. 97, p. 254). The weaver ants have shown us quite a different method of weaving a nest by using the spinning silk of their larvae (figs. 48 and 49, pp. 112 and 113).

The skillfully woven spherical home of an African

penduline titmouse is even equipped with a door that closes across the entrance hole when the bird is not in the nest. Closing devices exist with a number of animals, but the solutions of the problem could hardly be more varied. The trap-door spiders make lids for their soil tubes (fig. 17, p. 35, and pl. 19, p. 38), and some ants employ special doorkeepers that block the entrances with their own heads and only admit members of their own colony (fig. 43, p. 103). Some solitary bees block the entrances to their homes in a similar way, by sitting in the opening when they are alone in their nests.

These are very primitive methods. But the hinged lid of the trap-door spider anticipates a technical construction which man did not invent until much later. Such "anticipated inventions" without an inventive mind are legion: the ant lion digs a pit into the ground (fig. 12, p. 25) much as our primeval ancestors did to trap their prey. Spiders spin their webs to ensnare their winged victims in the manner of bird catchers (pl. 14, p. 37). A caddis fly larva catches its food with a funnel-shaped fishing net similar to those used by certain fishermen (fig. 18, p. 42). We recall the masonry of mason bees (fig. 30, p. 70, and pl. 40, p. 78), and the well-digger jawfishes (fig. 71, p. 164); the pottery of the potter wasps (pl. 27, p. 57), of wasps of the genus Polybia (pl. 37 a, p. 77), and of the ovenbirds (fig. 84, p. 226). Among the most remarkable of these inventions are the ventilation systems in the mounds of certain termites (figs. 60 and 61, pp. 140 and 142), built by minute workers active in a tiny part of a large building and yet cooperating in the construction of a harmonious whole without a blueprint and without directions from an architect.

We humans are proud of our inventions. But can we discern greater merit in our capabilities than in those of the master builders who unconsciously follow their instincts? The evolutionary roots of human behavior reach far back into the behavior patterns of animals. Those who are fascinated by these connections need only fasten on one such puzzle, the architecture of animals perhaps, to find an absorbing interest for a lifetime. Gradually they may learn to understand a great deal that at first sight appeared incomprehensible, and to people of an inquiring cast of mind, this will afford deep satisfaction. And yet the sum total of unsolved mysteries will always remain immeasurably greater than the sum of our discoveries.

There are biologists who are convinced that they, or

future generations of scientists will ultimately find the key to life in all its manifestations, if only research perseveres. They are to be pitied. For they have never experienced that sense of profound awe in the face of the workings of nature, some of which will forever elude comprehension, even by the mind of man.

Table of Metric Equivalents

Unit		Metric Equivalent			U.S. Equivalent	
millimeter	=	0.001	meter	=	0.03937	inch
centimeter	=	0.01	meter	=	0.3937	inch
meter	=	1.0	meter	=	39.37	inches
kilometer	=	1000.0	meters	=	0.6214	mile
gram	=	1.0	gram	=	15.43	grains
kilogram	=	1000.0	grams	=	2.2	pounds

Picture Sources

Drawings | Figures

The drawings throughout were made by Turid Hölldobler, most of them especially for this volume. Those listed below were prepared from the following sources:

11. Ant lion. Hesse-Doflein, *Tierbau und Tierleben,* Jena, 1943.
13. Spider's spinning glands. H. M. Peters, in *Zeitschrift für Naturforschung* 10 b, Tübingen, 1955.
19 b. Caddis fly and larva. Wesenberg-Lund, *Biologie der Süsswasserinsekten,* Berlin and Vienna, 1943.
21. Casings of caddis fly larvae and bagworms. 1, 3, 4 from W. Engelhardt, *Was lebt in Tümpel, Bach und Weiher?* Stuttgart, 1962; 2, 5, 6, 7 from Wesenberg-Lund, *ibid.;* 11, 12 from B. Grzimek, *Tierleben* II, Zurich, 1969; 13 from W. Dierl, in *Kumbu Himalaya* 4, Innsbruck-Munich, 1971.
24 b. Diagram of digger wasp with caterpillar. G. B. Baerends, in *Tijdsschrift voor Entomologie* 84, 1941.
25. Nests of potter wasp. H. Bischoff, *Biologie der Hymenopteren,* Berlin, 1927.
39. Honeybees' irregular cell construction. Martin and Lindauer, in *Zeitschrift für vergleichende Physiologie* 53, Berlin-Heidelberg-New York, 1966.
40. *ibid.*
43. Ant doorkeeper. E. O. Wilson, *The Insect Societies,* Cambridge, Mass., 1966.
45. Diagram of ants' ground nest. B. Gray, in *Insectes sociaux,* vol. 18, number 2, Paris, 1971.
56. Diagram of ground nest of termites. P. E. Howse, *Termites, a study in social behaviour,* London, 1970.

60. Diagram of termites' nest with fungus gardens. M. Lüscher, in *Acta Tropica* 12, No. 4, Basel, 1955 (altered).
61. *ibid.*
62. Termites' nest. P. E. Howse, *ibid.*
63. Diagram of termites' water shaft. P.-P. Grassé, *Traité de Zoologie* 9, 1949.
65. Diagram of termites' building technique. P. E. Howse, *ibid.*
66. Termites building arch. E. O. Wilson, *ibid.*
68. Nest of sand goby. F. Guitel in W. Wunder, *Ergebnisse der Biologie,* Berlin, 1931.
71. "Well-digger" jawfish. *Urania Tierreich,* Frankfurt-Zurich, 1969.
89. Bowerbird. Heinz Sielmann, *Lockende Wildnis,* Gütersloh and Vienna, 1970.
92. Unsuccessful nests on ladders. O. Heinroth, *Aus dem Leben der Vögel,* Heidelberg, 1955.
93. Upper right, mole, tunneling. O. Schmeil, *Lehrbuch der Zoologie,* Heidelberg, 1950; bottom left, claw of mole cricket. Hesse-Doflein, *ibid.* II.
96. Mammal teeth. Hesse-Doflein, *ibid.* I.
97. Harvest mouse nest. Sigbert Mehl and Herman Kahmann, *Einheimische Kleinsäugetiere,* Munich, 1963.
104. Chimpanzee in sleeping nest. Jane van Lawick-Goodall, *My Friends the Wild Chimpanzees,* Washington, D.C., 1967.
105. Chimpanzee with packing crates. Wolfgang Köhler, *The Mentality of Apes,* London, 1925.

Photographs | Plates

1. Nummulitic limestone. Bayerische Staatssammlung for Paleontology and historical Geology, Munich. Photo: Christa Schulz.
2. Tube worm *Megalomma*. Dr. Frieder Sauer, Munich.
3. *Euplectella aspergillum*. Dr. Max Renner, Munich.
4. Skeletons of Radiolaria. Dr. Frieder Sauer, Munich.
5 and 6. Coral polyps. Peter Kopp, Munich.
7. Coral skeleton. Elsa Grube (Dr. Werner Wrage), Amsterdam.
8. Coral polyps under water. Helmut and Günther Fleissner, Frankfurt.
9. Barrier reef. Dr. Werner Wrage, Hamburg.
10. Atoll. V-DIA-Verlag, Heidelberg.
11. Coral reef. Dr. Irenäus Eibl-Eibesfeldt, Seewiesen.
12. Snail shells. Bayerische Zoologische Staatssammlung, Munich. Photo: Dr. Max Renner.
13. Snail with bell animalcules. Dr. Frieder Sauer, Munich.
14. Spider's web. Alfred Limbrunner, Dachau.
15. Spider's thread. Dr. H. M. Peters, Tübingen.
16. White "fairy lamp." Dr. Max Renner, Munich.
17. Fairy lamp covered by soil. Ullstein GmbH Bilderdienst, Berlin. Photo: Dr. König.
18. Water spider. Dr. Frieder Sauer, Munich.
19 a, b, c. Trap-door spider. Dr. Friedrich Schremmer, Heidelberg.
20. Caddis fly larva. Dr. Frieder Sauer, Munich.
21. Leaf mines. Dr. Max Renner, Munich.
22. Bagworm crawling. Ullstein Bilderdienst, Berlin. Photo: Dr. König.
23. Bagworm on stalk. Dr. Frieder Sauer, Munich.
24. Lacewing eggs. Dr. Frieder Sauer, Munich.
25. Cuckoo spit insect larva. Zentrale Farbbildagentur, Düsseldorf.
26. Digger wasp. Dr. Frieder Sauer, Munich.
27. Potter wasp's nests. Private collection, Brunnwinkl. Photo: Dr. Max Renner, Munich.
28 and 29. Potter wasp and nest. Günther Olberg, Niemegk, German Democratic Republic.
30. Sand wasp. Günther Olberg, Niemegk, German Democratic Republic.
31 a, b. Solitary wasp's tube. Dr. Frieder Sauer, Munich.
32 a, b. Young wasps' nest. Dr. Max Renner, Munich.
33. Hornets' nest. Dr. Karl von Frisch, Munich.
34. Field wasp's nest. Dr. Frieder Sauer, Munich.
35. Wasps' nest. Dr. Frieder Sauer, Munich.
36. Wasps' nest. Okapia, Frankfurt. Photo: Kratz
37 a, b. Wasps' nest made of clay. Dr. Friedrich Schremmer, Heidelberg. Photo: Dr. Max Renner.
38. Paper nest of wasp *Chartergus*. Dr. Friedrich Schremmer, Heidelberg.
39. Ceramic nest of *Polybia singularis*. Dr. Friedrich Schremmer, Heidelberg.
40 a, b. Mason bee's nest. Dr. Max Renner, Munich.
41. Honeybees. Dr. Fritz Leuenberger, Bern.
42. European carder bees. Private collection, Brunnwinkl. Photo: Dr. Max Renner, Munich.
43. Bumblebees' nest with chicken feathers. Private collection, Brunnwinkl. Photo: Dr. Max Renner, Munich.
44. Comb of stone bumblebees. Dr. Frieder Sauer, Munich.
45. Brood nest of honeybees. Dr. Karl von Frisch, Munich.
46. Wooden beehive. Dr. Karl von Frisch, Munich.

47. Bee cells. Dr. Friedrich Ruttner, Oberursel.
48. Comb in construction. Dr. Friedrich Ruttner, Oberursel.
49. Irregular bee comb. Dr. Martin Lindauer, Würzburg.
50. Bee with pollen. Dr. Friedrich Ruttner, Oberursel.
51. Bee with propolis. Dr. Friedrich Ruttner, Oberursel.
52. Dwarf honeybees. Dr. N. Koeniger, Oberursel.
53. Anthill of *Formica polyctena*. Dr. Bert Hölldobler, Cambridge, Mass.
54. Anthill cross section. Model in the Naturhistorisches Museum der Technischen Universität, Braunschweig. Photo: G. Linhardt.
55 a, b. Larch stem. Specimen in the Naturhistorisches Museum, Braunschweig. Photo: Dr. A. Brauns.
56. Carton nest. Drawing by Turid Hölldobler, Cambridge, Mass.
57. Carton nest. Drawing by Turid Hölldobler, Cambridge, Mass.
58 a, b. Structures of compass termites. The Australian Government Information Office.
59. Nest of *Macrotermes carbonarius*. Dr. Karl Daumer, Munich.
60. Termitary with chimneys. Heinz Sielmann, Munich.
61. Termitary. Heinz Sielmann, Munich.
62. Termites' mound with rain roofs. Dr. Werner Wrage, Hamburg.
63. Mushroom-shaped termitaries. Dr. Martin Lüscher, Bern.
64. Anteater and tree termites' nest. Okapia, Frankfurt. Photo: A. Root.
65. Termite gallery. Dr. Martin Lüscher, Bern.
66. Paradise fish. H. J. Richter, Leipzig.
67. Fighting fishes. R. Zukal, Brno, Czechoslovak Socialist Republic.
68 a, b. Stickleback nest-building. Ziesler, Munich.
69. Leatherback turtle. Eugen Schuhmacher, Munich.
70. Brush turkey chick. Heinz Sielmann, Munich.

71. Brush turkey. Eugen Schuhmacher, Munich.
72. Thermometer fowl. Heinz Sielmann, Munich.
73. Nest of ringed plover. A. Limbrunner, Dachau.
74 a, b. Eider duck and nest. Dr. G. Bergman, Helsinki.
75. Swans and nest. Ullstein Bilderdienst, Berlin.
76. Least bittern and nest. Günther Olberg, Niemegk, German Democratic Republic.
77. Osprey eyrie. Ullstein Bilderdienst, Berlin. Photo: R. Siegel.
78. Nest of horned grebe. Eugen Schuhmacher, Munich.
79 a, b. Eagle eyrie. Dr. Franz Niederwolfsgruber, Innsbruck.
80. Great reed warblers' nest. A. Limbrunner, Dachau.
81. Reed warblers and nest. Dr. Otto von Frisch, Braunschweig.
82. Brush turkey on mound. Heinz Sielmann, Munich.
83. Lesser whitethroat brooding. Dr. Otto von Frisch, Braunschweig.
84. Leaf warbler and nest. Dr. Otto von Frisch, Braunschweig.
85. Wrens' nest. Ullstein Bilderdienst, Berlin. Photo: Siegel.
86. Storks' nest. A. Limbrunner, Dachau.
87. Penduline titmouse and nest. Okapia, Frankfurt. Photo: Dr. Sauer.
88. Nest of penduline titmice. Photo: G. Linhardt, Braunschweig.
89. Tree with nests of weaverbirds. Dr. Werner Wrage, Hamburg.
90 a, b. Textor weaverbird. Dr. K. Immelmann, Bielefeld.
91 a, b. Nest of weaverbird *Malimbus cassini*, and enlarged detail in natural size. Naturhistorisches Museum, Braunschweig. Photo: G. Linhardt.
92. Nests of oryx weaverbirds. Photo: Okapia, Frankfurt.
93 a. Nest of sociable weaverbirds. Eugen Schuhmacher, Munich.
93 b. Colony of sociable weaverbirds. Zentrale Farbbildagentur, Düsseldorf.

94. Great spotted woodpecker. Ullstein Bilderdienst, Berlin. Photo: Siegel.
95 a, b. Black woodpecker feeding young. Heinz Sielmann, Munich.
96. Bowerbird with red ornament. Heinz Sielmann, Munich.
97. Satin bowerbird. Heinz Sielmann, Munich.
98. Lauterbach bowerbird. Heinz Sielmann, Munich.
99. Painting of bowerbirds by L. Binder, based on color photographs and descriptions of Heinz Sielmann, Munich. Photo: Dr. Max Renner, Munich.
100. Bicycle shed. Dr. Otto von Frisch, Wendeburg.
101 a, b. Harvest mouse and nest. Dr. Lieselotte Dorfmüller, Munich.
102. Nest of dormouse. Dr. Otto von Frisch, Braunschweig.
103. Beaver dam. Courtesy of the American Museum of Natural History, New York.
104. Beaver dam and lodge. Dr. Alfred Bailey, Denver.
105. Beaver swimming. Ullstein Bilderdienst, Berlin.
106. Beaver dam. Eugen Schuhmacher, Munich.
107. Beaver lodge. Ullstein Bilderdienst, Berlin.
108. Beaver felling tree. Courtesy of the American Museum of Natural History, New York.
109 a, b. Tree gnawed by beavers. Eugen Schuhmacher, Munich.
110. Badger. Zentrale Farbbildagentur, Düsseldorf.
111. Squirrels' nest. A. Limbrunner, Dachau.
112 a, b. Marmots. A. Limbrunner, Dachau.
113. Beaver repairing dam. Courtesy of American Museum of Natural History, New York.
114. Beaver carrying bough. National Audubon Society. Photo: © Leonard Lee Rue III.

Subject Index

Aardvarks, 131

Africa, 128, 140, 142, 144, 163, 184, 192, 204, 207, 211, 212, 221, 223, 224, 227, 228, 230

Alligators, 170

Alps, 13, 141, 251, 263

Amoeba, 3–4, *3;* feeding method, 3–4; movement, 3–4; propagation, 4

Amphibia, 23, 151, 165–69, 211; brood care, 165; egg-laying, 165; larvae, 165; nests, 165–68; *see also* Frogs; individual animals

Ant lions, 25–27, 30; egg-laying, 25; feeding method, 25–26; growth, 26; innate drives, 26, 27; as larvae, 25–27; pit traps of, 25–26, *25, 284, 286;* poison of, 25

Antartic, 231–32

Anteaters, 131, *136*

Anthills, *see* Ants, mounds of

Ants: anatomy, 101, 103; and ant lions, 25–26; and aphids, 105, 111, 119–20, *119, 122;* and birds, 217; brood-raising, 104–05, 106–07, 109, 115, 123; use of camouflage, 103, 115; castes, 101, 103; and coccids, 105, 111, 119–21, *120;* cocoons of larvae, 104, 111, 115; colonies, 103–04, 106, 123; communication among, 103; doorkeepers, 103, *103,* 110, 286; egg-laying, 104, 123; feeding method, 104, 105, 111, 116, 118, *118,* 119–21, *119,* 123; females, 101, *102,* 103; hibernation, 105, 110; as hunters, 49, 119, 123; larvae, 104–05, 111–15, 125; life span, 104; males, 101, *102,* 103; nests, 106–19, 284, 285; nests, chambers of, 115–19; nests, defense of, 105, 118; nests, entrance holes of, 103, *103,* 107; nests, temperature of, 105, 108–09; numbers of, 101, 105–06, 107, 119; poison of, 105, 112; queens, 103–04, 106, 119, 121, 123; reproduction, 101, 103, 104; roads of, 22, 122; sexual, 101, 103; size differentia-

tion, 101, 117; soldiers, 101, *102,* 103, 117; storage chambers, 115–16; swarming, 103, 121; vagrant, 122–23; webs, 111–15, *112, 113;* workers, 101, *102,* 104, 105, 106–09, 116, 117–18, 120, 127; *see also* Ants, mounds of; Tree cavities; individual ants

Ants, carpenter, *99,* 110, 128

Ants, desert: workers as storage vats, 116, *116*

Ants, garden, 120; brown garden, 96, 104, *105;* mounds of, 96, 109, *109*

Ants, harvester, 115

Ants, jet; carton nest, *100,* 110–11, 285; division of labor, 111; fungus cultivation, 110–11, 119

Ants, leaf-cutter, *102,* 104, *117, 118,* 122; fungus cultivation, 116, 118–19, *118,* 127, 137; numbers of, 119; parasitic flies attack of, 117–18; queens, 119, 121–22; workers, 117–18

Ants, meadow: mounds of, 96, 109; roads of, 122; yellow, 96

Ants, mounds (anthills) of, 96, *98,* 107–10, *109;* angle of, 108–09, *109;* building materials, 96, 107, 109; constant reconstruction, 108; construction, 107–08; entrance holes, 107, 108; heating, 108–09, *109;* mold prevention, 108; size, 107, 109; *see also* individual ants

Ant, weavers, 120; use of camouflage, 115; larvae, as spinners, 111–15, *113,* 285; web nests, 111–15, *112, 113,* 285

Ants, wood, 101, *106,* 109; mounds, 96, *98,* 107, 108; numbers of, 107; red, 96, *102*

Apes, 278–82

Aphids, 105, 111; symbiotic relationship between ants and, 119–20, *119,* 122

Argentina, 167

Arthropods, 23, 151, 223; *see also* individual animals

Asia, 107, 110, 111, 128, 157, 164, 170,

185, 203, 204, 207, 221, 224, 227, 228, 251, 254, 267

Atolls, 12, 13, *13, 19*

Auks, 183, 184

Australia, 106, 107, 138, 172, 177, 179, 192, 237, 238, 239

Badgers, 251–53; feeding method, 251, 253; subterranean burrows, 251–53, *252,* 262, *272,* 284

Baerends, G. P., 51–52

Bagworms or case worms, *39,* 47–49, *48, 49; see also* Moths, psychid

Beavers, *253,* 266–68, 273–78; artificial islands of, 268, 273; brain, 277; colonies, 267, 275; dams, 22, 248, *266, 267, 270,* 273–75, *273, 274;* feeding method, 277; hands, 276, *276;* innate drives, 278; life span, 277; litters, 277; lodges, 247–48, *267,* 268, *268, 270,* 273, 274; teeth, *271,* 275, 276; transportation of wood, *275, 276;* water-level regulation, 268, *268, 274;* as woodcutters, *271,* 275–76

Becker, Günther, 139

Bee-bread, 67, 70

Beekeeping, 81–82, 96; hives, *80,* 81–82, 85, 88, 93, 95; and swarming, 84–85

Bees, 66, 101; *see also* Bumblebees; Honeybees; individual bees

Bees, carder, 73, *78;* brood sizes, 74–75; building material, 73, 74, 75; cocoons of larvae, 74; egg-laying, 73; feeding method, 73, 74–75; males, 75; nests, 73–75, *74;* queens, 73–75; storage cells, 74; wax of, 73; workers, 74

Bees, mason, 68–71; building material, 69–71, *70, 78;* use of camouflage, 70; egg-laying, 69, 70, *70;* feeding method, 69, 70, *70;* flight hole closures, 69, *69,* 70; innate drives, 69; insulation of nests, 69; leaf-cutter, 68–69, *69;* nests, 69–71, *69, 71, 78,* 286

Bees, mining: communal life, 72; females, 72; four-banded, 72; hibernation, 72; males, 72; nests, ventilation system for, 68, *68;* workers, 72

Bees, plasterer or gum, 67; feeding method, 67; nests, 67; waterproofing nest, 67, 68

Bees, social, 66, 72–96; *see also* Bumblebees; Honeybees; individual bees

Bees, solitary, 51, 66–72; anatomy, 67; beebread, 67; cuckoo bees, 72; feeding method, 67; females, 51; larvae, 67–68; males, 51; mold as danger to, 68; nests, 66–72, 284, 286; parasites of, 72; transition to social bees, 72; ventilation system, 68; wax of, 68; *see also* individual bees

Bees, yellow-faced, 67; feeding method, 67; nests, 67; waterproofing nests, 67

Birds, 151, 171–72, 177–244; aesthetic feelings of, 244, 246–47; breeding sites, 183–84, 188; brood, 187–88, *187;* brood care, 189; brooding, 185, 191; cave-breeding, 217; colonies, 192, 207, 211, *212, 215, 216,* 228, 232; courtship, 188, 237; egg-laying, 185; females, 188–89, 191; hole-breeding, 218–23; innate drives, 152, 188, 189, 190, 210–11, 244–45, 247; and insects, 217, *218;* males, 188–89, 191; mental activities of, 23, 244–47, 284; migratory, 188; nesting materials, 189–90, 197, *201;* nests, 151, 184–85, 187–231 *passim,* 245–46; nidicolous, 187, *187,* 191; nidifugous, 178, 187, *187,* 191; plumage, 172, 237; and reptiles, 171, 182; soil-breeding, 223–25; songs of, 188, 219, 237; stealing of, 90–91; temperature of, 171–72; as tenants, 217; tools of, 189, 218, 238; young, nests of, 210–11; see also Megapodes; Mounds, brooding; Nests; individual birds

Bitterns, 192; brooding, 192; use of camouflage, 192; females, 192; least, *173,* 192; males, 192; nests, 192

Blackbirds: European, 245–46; nests, 245–46, *246, 247*

Bowerbirds, 237–42; esthetic feelings of, 244, 246–47; bowers, *234, 235,* 237–42, *242,* 284; courtship, *235,* 237–42; dance of, 237, 238, 239, 240, *241,* 242; females, 237, 238, 239, 240, 242; great, *234,* 239, *241;* Lauterbach, *235,* 240–41, *242,* 243; maintenance of bower, 238–39; males, 237–42; mating, 239, 240; nests, 237; plumage, 238, 239, 240; satin, *235,* 238–39, 241, *242;* song of, 239; use of tools, 238; young, 239; *see also* Orange-crested gardener

Brazil, 167

Britain, 248, 251, 262, 267

Brunnwinkl, 75
Brush turkey, 177–78; brood, 177–78; brooding mound, *154, 177, 178;* cocks, 177; egg-laying, 177, 178; hens, 177; temperature of mound, 177
Building activities, 22–23, 238–87; and aesthetic feelings, 244–45, 246; division of labor, 83–84, 85, 101; and innate drives, 23, 26, 54, 63, 69, 82, 89, 151–52, 211, 225, 244–45, 247, 278, 280, 282, 283–84, 286–87; and mental activities, 23, 149, 244–47, 281–82, 284; and protection of brood, 34, 43, 47, 49–50, 52, 66, 68–71, 81, 105, 129, 168, 169, 284–285; variety, 22, 50, 66–67, 284; *see also* individual animals
Bumblebees, 66, 72–76, 81; brood sizes, 74–75; building materials, 73, 75–76, *79;* combs, 74–75, 76, 86; feeding method, 73–75, 81; hibernation, 73, 81; life span, 81; males, 75; nests, 73–76, *76, 79,* 96; nests, covers for, 75–76; queens, 73–75, 81, 83, 104; stone, 76, *76, 79;* storage cells, 74; wax of, 73, 76; workers, 74, 127; *see also* Bees, carder
Butterflies, 45

Caddis flies, 41, *44,* 45
Caddis flies, larvae of: building materials, 43, 44; use of camouflage, 43; casings, *39;* 41, 42–45, *44, 45,* 47, *48;* feeding method, 44; underwater nets, 41, 42, *42,* 43, 284, 286
Camouflage: birds, 192; casings, 43, 47; cocoons, 34; eggs, 184, 185; nests, 65, 70, 103; tubes, 36
Carps, 157
Casings: bagworms, 47–48, *48, 49;* caddis fly larvae,. *39,* 41, 42–45, *44, 45,* 47, *48;* camouflage of, 43, 47; *Difflugia,* 4; Foraminifera, 4; *see also* Cocoons
Caterpillars, 41, 43; of clothes moth, 46–47, *47;* feeding method, 46–47; leaf-mining, 46; as prey of wasps, 51, 52–53, *52, 53,* 54, 55, *57;* of psychid moths, 47–49, *49*
Central America, 185, 225, 230
Ceylon, 114, 120, 131, 145–46
Chile, 168
Chimpanzees, 278–82; aesthetic feelings of, 244–45; innate drives, 280, 282; mental activities of, 281–82, *281;* sleeping nests, 278–80, *279,* 282; young of, 280
China, 227, 228
Cichlids, 163: as mouthbrooders, 164–65
Clay nests, 55–56, *57,* 64, 89, 167–68, 225–26, 286
Coccids, 105, 111; symbiotic relationship between ants and, 119–21, *120*
Cocoons: of ants, 104, 111, 115; of bees, 74; camouflage of, 34; egg, of spiders, 34, 36; insect, 41: *see also* Casings
Collias, N. E., 190
Colombia, 64, 115
Combs: of bumblebees, 74–75, 76, 86; movable, 82; ornamentation of, 89; of wasps, 61, 62, 85; wax, construction of, *80, 85–95, 86, 87, 91, 92, 93, 97;* wax, of honeybees, 61, *79, 80,* 81–82, 83, 84, 85, *86, 88,* 96
Coral polyps, 10–14, *11, 18;* atolls, 12, 13, *13, 19;* dangers to, 11–12; dividing of, 10–11, *18;* fossils, 13; growth, 10, 12–13; larvae, 10–11; reefs, 11–13; *19,* 283; skeletons, 10–13, 21, 283; *see also* Polyps
Corvids, 188
Cranes, 191
Crocodiles, 169, 170; brood care, 170; defense of nests, 170–71; egg-laying, 170; females, 170–71; nests, 170–71; saltwater, 170–71; temperature of nests, 171, 172
Crows, 191
Cuckoo spit, *see* Froghopper
Cunners, 163 *n.*
Cyclops, 9

Daphnia, 9
Darwin, Charles, 12, 168
Difflugia, 4, *4;* casings, 4; feeding method, 4; propagation, 4
Doflein, Franz, 114, 120
Dormice, common, 247, 258; hibernation, 258; nests, 247, 258, *258*
Dotterels, 191
Ducks, 189, 191; *see also* individual species

Eagles, 189, 217; bald, 197; golden, *175,* 197, 265
Eibl-Eibesfeldt, Dr. Irenäus, 117

Eider ducks, 185; down-padded nests, *156,* 185

Escherich, Karl, 131, 145–46

Ethiopia, 143

Europe, 44, 96, 107, 115, 128, 157, 161, 185, 192, 198, 201, 203, 204, 224, 227, 228, 248, 251, 254, 259, 267

Excreta as building material: birds, 222; termites, 132, *133, 143,* 144, 146, 147–48, *148,* 285

Eyries, *174, 175,* 189, *195,* 197, 217

Fairy lamp, *see* Spiders, European tube

Falcons, pygmy, 217

Finches, 191

Fishes, 23, 151, 152–65, 168, 188, 211; and frogs, 168; nests, 152, 157, 158–64; *see also* individual fishes

Fishes, fighting, *153,* 157, 159, 160

Flickers, 190; courtship, 221; yellow-shafted, 221

Flies, 23; parasitic, 117–18; as spiders' victims, 28, 29

Flycatchers, 191

Foraminifera, 4–6, *5;* casings, 4–6, *6, 7;* propagation, 4–5, 7; skeletons, 6, 9

Fossils, 5, 13

Foxes, 252

France, 128

Frith, H. J., 180–81

Froghoppers, 49–50; cuckoo spit foam nest, *40,* 49–50, *50,* 165, 285

Frogs, 165–69; clay nests, 167–68; and fishes, 168; flying, 167; foam nests, 165–67; *see also* Tadpoles; individual frogs

Frogs, African tree: brood care, 167; females, 167; foam nests, 165, 167; tadpoles, 167

Frogs, Darwin, 168–69, brood care, 169; egg-laying, 168, 169; females, 168; males, 168–69

Frogs, edible: egg-laying, 169

Frogs, Javanese flying: egg-laying, 166; females, 166–67; foam nests, 165, 166–67, *166,* 285; males, 166; mating, 166, *166;* tadpoles, *166,* 167

Frogs, Omai "rowing,": foam nests, 165

Frogs, tree, 167–68; clay nests, 167–68, *168;* males, 167–68; mating, 168; tadpoles, 168

Fungus cultivation: of ants, 110–11, 116, 118–19, *118;* of termites, 128, 132, *133,* 137, 139, *140,* 141

Germany, 204

Gombe Stream Chimpanzee Reserve, 278

Goodall, Jane, 278–80; *In the Shadow of Man,* 278

Goodwin, A. P., 242–43

Gorillas: sleeping nests, 280

Grassé, Pierre-P., 137

Greenland, 183

Griffin, Donald R., 117

Guillemots, 183; egg-laying, 183

Gulls, 185

Haeckel, Ernst: *Kunstformen der Natur,* 6; *Radiolaria,* 6

Hamburg, 128

Hamsters, 266

Harvest mice, European, 247, 254–55, *256,* 266; feeding method, 253; nests, 247, 254–55, 285; nests, breeding, *254,* 255, *257*

Heinroth, Oskar, 186

Hermit crabs, 21

Herons, 189; colonies, 192; gray, 192; males, 192; nests, 192

Hibernation, 253; ants, 105, 110; bees, 72, 73, 81; dormice, 258; marmots, 263–64; snails, 16; wasps, 56

Hives, *80,* 81–82, 85; temperature of, 95; *see also* Combs

Hölldobler, Professor Bert, 110

Hölldobler, Turid, 114

Honey: as food for bees, 69, 83, 85; stored by bees, 73, *79,* 81, 85, 87, 93, 96

Honeybees, 66, 81–96, 105; anatomy, 67, 83, 87, *88,* 89–92, *90, 91, 92, 97;* breeding cells, 93; brood feed, 83; colonies, 82–85, 96; communication among, 84, 85; division of labor, 83–84, 85; drones, *78,* 83, 85, 89, 91; dwarf, 95, *97;* egg-laying, 85, 91; feeding method, 83, 85; hives, *80,* 81–82, 85; innate drives, 82, 89, 283–84; life span, 83, 84; measuring instruments of, 89–92, *90;* numbers of, 82–83; orientation of, by magnetic field, 93–94, *93;* ornamentation of combs, 89; poison of, 81; pollen and honey storage,

79, 81, 85, 87, 93, 96; pollen and nectar gathering, 83, 84, 94, 97; propolis, gathering and use of, 94–95, 97; queens, 78, 83, 84, 85, 86, 91, 96, 103; reproduction, 83, 91; sex determination, 91; stingless, 96; swarming, 84–85, 93–94, 96, 103; temperature of hive, 95; wax combs, 61, 79, 80, 81–82, 83, 84, 85, 86, 88, 96; wax combs, construction of, 80, 85–95, 86, 87, 91, 92, 93, 97; workers, 78, 83–85, 89, 91, 95, 96, 103, 127

Hornbills 221–23; beaks, 221–22; breeding cavity, 222–23, 222; brood, 222–23; feeding method, 222, 223; females, 222; males, 222; molting, 222, 223

Horned grebe, 174, 192

Hornets, see Wasps, social true

Howse, P. E., 149

Hummingbirds, 191, 198–201, 228; females, 190, 191, 199–200; males, 199; nesting materials, 199; nests, 199–201; stealing of, 190, 245; temperature of, 199, 200; violet-eared, 190; white-eared, 190, 199, 201

Hungary, 197

Iceland, 185

India, 95, 111, 227, 230

Indonesia, 228

Insects, 23, 24, 30, 151; and birds, 217, 218; cocoons, 41; egg-laying, 50; innate drives, 151, 211; larvae, as prey, 218–19, 250; larvae, spinning glands of, 41, 111–15; 285; mental activities of, 149; social, 22, 55, 65–66, 72–73, 101, 106; trappers, 22, 24–41, 284, 286; see also individual species

Japan, 43, 251, 254

Jawfishes, 164; anatomy, 164; as mouth-brooders, 164 and n., 165; nests, 164, 286; well-digger, 164, 164

Jefferson River, 273–74

Kingfishers, 223–25; courtship, 224; European, 223–25; feeding method, 224; males, 224; soil-breeders, 223, 224–25, 284

Kirchner, Dr. Georg, 86–87

Kloft, Werner, 108

Koeniger, Dr. N., 95

Köhler, Wolfgang, 280–81; Mentality of Apes, The, 281

Kutter, Henri, 104

Labyrinth fishes, 157–59; bubble nests, 153, 157, 158–59, 158, 165–66, 285; colors, 158, 159; defense of nest, 158; females, 158–59, fertilization, 158–59; fighting fishes, 153, 157, 159, 160; males, 157–59; see also Paradise fishes

Lacewings, 50; egg-laying, 40, 50

Larvae: insect, as prey, 218–19, 250; insect, spinning glands of, 41, 111–15, 285; protection of, 34, 43, 47, 49–50, 52, 66, 68–71, 81, 105, 129, 168, 169; see also Caterpillars; individual animals

Leaf: -cutter ants, 102, 104, 116–19, 117, 118, 121–22, 127, 137; -cutter bees, 68–69, 69; -mining caterpillars, 46

Lindauer, Professor Martin, 93

Lizards, 169, 171; monitor, 170

Lovebirds, 189–90

Lüscher, Professor Martin, 140, 143, 146

Madagascar, 111, 230

Magpies, 203

Mallard ducks, 185; brood care, 186–87; females, 185–87; innate drives, 186, 245; nests, 185–86

Mallee birds (thermometer fowl), 179–82; brooding mound, 155, 179–82; cocks, 179, 181; egg-laying, 179, 181–82; hens, 179, 181; temperature of mound, 155, 179–81; ventilation of mound, 179

Mammals, 23, 151, 171, 188, 223, 247, 284; see also individual mammals

Marmots, 262–66; alpine, 262–65; colonies, 263; hibernation, 263–64; hoary, 265; as lookouts, 264–65, 264; mating, 264; as prey, 264; subterranean burrows, 262–66, 263, 272, 284; woodchucks, 265–66; yellow-bellied, 265

Martens, 251

Megapodes, 172, 177–83; brood, 181, 182; incubation by mounds, 155, 172, 177–82; incubation by sun, 182; incubation by volcanic heat, 182; innate drives, 183; see also individual animals

Mexico, 190, 199

Mississippi River basin, 273
Mole crickets, 248, *248*
Moles, 248–51; anatomy, 248, *248;* European, 248; feeding method, 250; food storage, 250–51; life span, 250; mating, 250, 251; molehills, 249; senses of, 250; subterranean existence of, 247, 248–51, *249;* ventilation system, 249
Molting: hornbills, 222, 223; trap-door spiders, 36
Moths, 45–49; clothes 46–47, *47;* egg-laying, 46, 47, 49; psychid, 47–49; reproduction, 47, 48–49; *see also* Caterpillars; individual moths
Mounds, *see* Ants, mounds of; Termites, mounds of
Mounds, brooding, *154, 178, 180;* crocodiles, 171; fermentation, 177, 179; megapodes, *155,* 172, 177–82; temperature, 177, 179–81; ventilation, 179
Munich Zoo, 282
Mushrooms, 137

Nectar: as food for bees, 66, 67, 69, 73; gathered by honeybees, 83, 84
Nepal, 47
Nests: camouflage of, 65, 70, 103, 202; carton, 110–11, 144, *145,* 285; clay, 55–56, *57,* 64, 89, 167–68, 225–26, 286; closures, 52–54, 69, 70, 73, 207, 286; communal, 211–12, 217; communications in, 64, 73; covers, outer, 75–76; cup-shaped, 163, 185, 191–92, 197–201, 204, 218, 230, 237; flight holes, 55, 61, 64–65, 69, 73, 212; flight tubes, 207, 209, 211, 224, *231;* foam, 49–50, *50, 153,* 157, 158–69, *158,* 165–67, 285; hanging, 204–10; insulation, 62, 69, 75, 137; paper, 56, *59, 60,* 61, 64–65, *77,* 89, 110–11, 144, 285; roofed, 202–04, 218, 255, 261; as roost, 202–04, 221; temperature, 62, 105, 108–09, 129, 138, 141, 171; underwater, 34–35, 159–64; ventilation systems, 68, 140–43, 149; waterproofing, 67, 68; of young animals, 210–11, 255, 280; *see also* Combs; Hives; Mounds; Tree cavities; Webs; individual animals
Nets: of caddis fly larvae, 41, 42, *42,* 43, 286; for trapping, 22, 41, 286

New Guinea, 172, 221, 237, 240–41, 242
North America, 44, 110, 111, 115, 116, 123, 128, 161, 170, 185, 191, 192, 199, 201, 204, 211, 221, 224, 225 *n.,* 230, 251, 253, 259, 262, 265, 267
Norway, 185
Nuthatch, European, 191

Oehmke, Dr. Martin, 93
Orange-crested gardener, 242–44; courtship, 243–44; dance of, 244; females, 244; huts (bowers), *236,* 242–44; maintenance of hut, 243–44; males, 242–44; mating, 244; plumage, *236,* 244; *see also* Bowerbirds
Orangutans: sleeping nests, 280, 282
Orioles, 191
Ospreys, *174,* 197
Ostriches, 184; cocks, 184, 197; egg-laying, 184; hens, 184
Ovenbirds, 225–26; clay nests, 225–26, *226,* 286; innate drives, 225
Owls, 191, 223

Pakistan, 95
Paper nests, 56, *59, 60,* 61, 64–65, *77,* 89, 110–11, 144, 285
Paradise fishes, 157–59; bubble nests, *153,* 157, 158–59, *158;* colors, 158, 159; defense of nest, 158; females, 158–59; males, 157–59, 160; mating, 158–59; *see also* Labyrinth fishes
Parrot fishes, 11–12
Parrots, 189, 217, 223, 284
Parthenogenesis, 48–49
Penduline titmice, 190; African, 206–07; females, 206; males, 204–06; nests, *196,* 204–07, *205, 206,* 218, 285–86
Penguins, 231–32, 237; breeding grounds, 232, 237; brood, 232, 237; brood pouch, 231, 232; colonies, 232; emperor, 231–32, *232;* feeding method, 232; females, 232; males, 232
Phalaropes, northern, 191
Pheasants, 189
Philippines, 230
Pigeons, 191
Plover, ringed, *156,* 184
Pollen: as building material, 73; as food for bees, 66, 67, 69, 73–74; gathered by

honeybees, 83, 84, 94, *97;* stored by honeybees, *79,* 85, 93, 96; transported by bees, 67

Polyps, 9–14; *see also* Coral polyps

Propolis, 94–95, *97*

Protozoa, 3–7; skeletons, 6–7, 10, 21, 283; *see also* individual animals

Pseudopodia, 3–4

Queens, *see* individual animals

Radiolaria, 6; propagation, 7; skeletons, 6–7, *8, 9, 17*

Reefs, coral, 11–13, 283; barrier, 13, *19;* fringing, 12, 13

Reproduction: amphibia, 166, 168; ants, 101, 103, 104, 124; beavers, 277; bees, 83, 91; coral polyps, 10–11; fishes, 152, 158–59, 160, 162, 163; large number of offspring, 50, 157, 169; parthenogenesis, 48–49; protozoa, 4–7; psychid moths, 47, 48; rodents, 251, 255, 261, 264; termites, 124–25, 127, 137; *see also* individual animals

Reptiles, 23, 151, 152, 169–71; and birds, 171, 182; brood care, 169; egg-laying, 169; flying, 171; *see also* individual animals

Rheas, 191

Ridley (scientist), 114

Rodents: as builders, 253, 266; teeth, 253–54, *253;* use of tools, 253; *see also* individual rodents

Russia, 267, 273

Ruttner, Professor Friedrich, 95

Salamanders, 165 and *n.;* larvae, 165

Salerno, 95

Saliva as building material, 285; bees, 69; birds, 222, 227–30, *229,* 285; labyrinth fishes, 158, 285; spiders, 35; termites, 144, 146, 285; wasps, 55, *58,* 61, 65, 285

Salmon, 152–53; egg-laying, 152, 157; females, 152; fertilization, 152; males, 152; migrations, 152; nests, 152

Sand goby, 159–60; egg-laying, 160; females, 160, males, 159–60; nests, 159–60, *160*

Schildknecht, Professor H., 118

Schremmer, Dr. Friedrich, 35, 64–65, 115

Scrub fowl, 178; brooding mound, 178

Sea anemones, 10

Sielmann, Heinz, 143, 220, 239, 240–41, 243

Silkworms, 41, 43

Skeletons: coral polyps, 10–13, 21, 283; protozoa, 6–7, *8,* 10, *17,* 21, 283; sponges, 9, 10, 21, 283

Snail shells, 14–16, *20,* 21; growth, 14–15, 21; as protection, 14, 16, 21; sealing, 16, *16,* 21; used by other animals, 21, 43, 70, *71;* variations, 15–16, *20*

Snails, 14–16, 21; anatomy, 14; hibernation, 16

Snakes, 169

South America, 96, 116, 122, 163, 191, 199, 225, 231

Sparrows, 217; European tree, 203; house (English), 203, 207 and *n.*

Spiders, 27–41; anatomy, 29, 30; use of camouflage, 34, 36; egg cocoons, 34, 36; innate drives, 151, 211, 283–84; memory of, 30; poison of, 30; spiderlings, 30, 36; spinning glands, 27–30, *27,* 41; *see also* Spiders' webs; individual spiders

Spiders, European tube, 34; egg cocoon, 34, *37*

Spiders, garden, 27–33, *28, 29;* feeding method, 30; females, 28; prey of, 29–30; sense of touch, 29, 33; spinning glands, 27–30, *27,* 41; *see also* Spiders' webs

Spiders, jumping, 34

Spiders, trap-door: egg cocoon, 36; feeding method, 36; life span, 36; molting, 36; tube hole, 35, *35,* 36, 286

Spiders, water, 34, *38;* air supply of, 34–35

Spiders, wolf, 34; egg cocoon, 34

Spiders' webs: construction, 30–34, *32;* as nesting material, 189, 198, 199–200, *201* 227; repair of, 33, *37;* as shelter, 28; silk of, 27, 189, 198, *201;* sticky spiral of, 28–29, *32,* 33, *37;* as trap, 28, 284, 286

Spittle bugs, *see* Froghoppers

Sponges, 9; skeletons, 9, 10, 21, 283

Squirrels, 259–62; American red, 262; feeding method, 253, 260; females, 261; food storage, 261–62; gray, 262; innate drives, 262, 282; nests (drays), 259, 260–61, *260, 266, 272;* red, 259, 262

Starfish, 12

Starlings, 190, 191, 217

Sticklebacks, 161–62; brood care, 162; colors of, 161; defense of nest, 161, 162; egg-laying, 161, 162; females, 161, 162; fertilization, 162; innate drives, 162; males, 161–62; nests, *153,* 161–62,, *161,* 285; three-spined, 161

Storks, 188, 189, 191, *195,* 217; European white, 197

Sugar: as building material, 74–75, *79;* as food for ants, 111, 119; and symbiosis between ants and aphids and coccids, 119–21, *119, 120*

Surinam, 121

Swallows, 188

Swan, mute, *173,* 187

Swifts, 228–31; colonies, 228; common, 228; lesser swallow-tailed, 230–31; nesting material, 227–28, 230; nests, 228–31, *229, 231;* palm, 230; saliva of, 228–30; swiftlets, 227–30

Tadpoles, 165; of African tree frogs, 167; of Darwin frogs, 169; of Javanese flying frogs, *166,* 167; of tree frogs, 168

Tailorbirds, 227, *227*

Tautogs, 163 *n.*

Tenerife Anthropoid Station, 281

Termites, 283–84; anatomy, 123–24, 125, 126–27, 149; and birds, 217; brood care, 125, 129, 137; building materials, 129, 144, 146, 147–48; castes, 126, *126,* 146; communication among, 139, 149; damage done by, 123, 127, 128, 129; and earth's magnetic field, 139; egg-laying, 125, *132;* feeding method, 127–28, 137; females, 124–25, 127; fungus cultivation, 128, 132, *133,* 137, 139, *140,* 141; kings, 125, 127, *132,* 137; larvae, 125–26, 137; life span, 125, 137; males, 124–25, 127; mental activities of, 149; nests, 124, 128–29, *129,* 131–50, *145;* numbers of, 123, 124, 140; passages and runways, 124, 128, 129, *133,* 137, 141, 145–47, *145, 147;* as prey, 125, 136; primitive species, 128; queens, 125, 127, 128, *132,* 137; reproduction, 124–25, 127, 137; roads of, 22, 146; royal cell, 125, *132, 133,* 137, 139, *140,* 141, 144, *145;* sense of smell, 146–47, 149; sexual, 124, *126,* 128; soldiers,

126, *126,* 127, 128, *132,* 146; swarming, 124, 128; water, need for, 143–44, *144;* workers, 125, 126, 127, 128, 129, *132,* 137, 144–45, *145,* 146, 147–48, 150; *see also* Termites, mounds of; individual termites

Termites, compass, *130, 131,* 138

Termites, desert, 144

Termites, mounds of, 129, 131–50, *135,* 284; angle of, 138; brood care, 129, 137; building materials, 129, *133, 143,* 144, 147–48, *148;* chambers, *129,* 131, *133, 135,* 137, 139, 141, 144; chimneys, *134,* 142–43; coloration, 132; communication in, 139, 149; fungus cultivation, 132, *133,* 137, 139, *140;* insulation, 137; layout, *133, 135,* 137, 139–41, *140,* 144, *144;* passages and runways, 124, 129, *133,* 137, 141; as protection, 129; roofs for, *135,* 138, *139;* size, 129, 131, 138, 139–40; temperature, 129, 138, 141; ventilation systems, 140–43, *142, 143,* 149, 286; water, need for, 143–44, *144;* workers, 129, 137, 147–48

Terns, 183, 185; breeding site, 184; fairy, 183; feeding method, 184

Tests, *see* Casings

Thailand, 159

Thermometer fowl, *see* Mallee birds

Thrushes, 188, 189

Tinbergen, Niko, 162

Titmice, 191, 204

Tools of animals, 22, 54, 55, 114, 189, 218, 238, 253

Tortoises, 169

Trappers, 22, 24–41, 284, 286; *see* also Ant lions; Spiders

Tree cavities as nesting sites: ants, 103, *103,* 110; birds, 218–21, 222–23; squirrels, 261

Trinidad, 117

Tubes: bristle worms, 16; camouflage of, 36; closures, 35–36, 56; polyps, 9; as protection, 16; spiders, 35–36; wasps, 56, *58; see also* Casings; Nests

Turtles, 169–70; egg-laying, 170, 182; enemies, of, 170; females, 170; leatherback, *154;* sea, 169

Uganda, 142

Underwater: nets, 41, 42, 43; nests, 34–35, 159–64; water spiders, 34–35
Unicellular animals, *see* Protozoa

Verdins, 206, 211
Vertebrates, 23, 151, 285; innate drives, 23, 284; lower, 23, 171–72; mental activities, 151, 284

Wagner, H. O., 190, 199
Warblers: blackcap, 198, 202; fan-tailed, 227; females, 198; great reed, *176,* 197; leaf, *194,* 202; lesser whitethroat, *193;* males, 198; nesting material, 197–98; nests, *194,* 217, *218;* reed, *176,* 197
Wasps, 23, 50–66; and birds, 217, *218;* combs, 61; females, 51; innate drives, 54, 63; males, 51; nests, 51–56, *59,* 61–65; protection of larvae, 50; social, 55, 56, 61–66; solitary, 51–56; *see also* individual wasps
Wasps, digger, 51–55, 67; brood, 53–55; egg-laying, 51, 53, *53,* 284; females, 51, 53, 55; innate drives, 54; larvae, feeding of, *40,* 51, 52–54, 284; males, 51, 55; nests, 51–55; prey paralyzed by, 51, 52–53, *52, 53,* 54; protection of nest, 52, *53, 58;* sand, 52–53, 55; sense of direction, 52; use of tools, 54
Wasps, paper: building materials, 56, 61, 285; combs, 61, 85; females, 56, 62; hibernation, 56; nests, 56, *59, 60,* 61, 64–65, *77,* 89, 110; queens, 56, 61, 62; *see also* Wasps, social true
Wasps, potter, 55–56; anatomy, 55; clay nest of, 55–56, *57,* 64, 89, 225, 286; egg-laying, 55–56, *56;* females, 55; larvae, feeding of, 55; prey paralyzed by, 55, *57;* use of tools, 55
Wasps, social true, 55, 56, 61–66; use of camouflage, 65; communication among, 64; community, 62, 65–66; division of labor, 62, 65; egg-laying, 62; females, 56, 62; innate drives, 63; insulation of nest, 62; larvae, feeding of, 62; life span, 64; nests, 56, *59, 60,* 61–65, *63, 77,* 286;

queens, 56, *60,* 61, 62, 63, 65–66, 83; temperature of breeding combs, 62; workers, 62, 63, 65, 127; *see also* Wasps, paper
Wasps, solitary true, 55–66; nesting tube, 56, *58; see also* Wasp, potter
Weaverbirds, 207–12, 217; buffalo, 211; Cassin's, 207–10, 211; colonies, 207, 211, *212, 215, 216;* communal nests, 211–12, 217; courtship, 210, 211, *213;* females, 208, 210, 211, 237; flight tubes, 207, 209, 211; males, 208–10, 211, 237; nests, 207–10, *208, 209, 213, 214, 216,* 218, 237, 284, 285; oryx, 211, *215;* sociable 211–12, *212, 216,* 217; textor, 211
Webs: of ants, 111–15, *112, 113;* construction of, 30–34, *32;* as nesting material, 189, 198, 199–200, *201,* 227; as shelter, 28; of spiders, 27–31, 33; as trap, 28, 284, 286
Well-diggers, 164, *164;* defense of nest, 164
Williams, R.M.C., 149
Wood rats, 258, 266; bushy-tailed (pack), 259; dusky-footed, 259; lodges, 258–59; white-throated, 259
Woodchucks, *see* Marmots
Woodpeckers, 203, 217, 218–21, 231; anatomy, 219, 220; black, 189, 220, *233;* brood, 220–21, *233;* communication among, 219; feeding method, 218–19, 220, *233;* females, 219, 220; food storage, 221; great spotted, 219, *233;* green, 221; hole-breeders, 218–21, *233;* holes, as roosts, 221; males, 219, 220; red-headed, 221
Workers, *see* individual animals
Worms: bristle, 16; earth-, 250–51; tube, 16, *17; see also* Silkworms
Wrasses, 162–63; defense of nest, 163; egg-laying, 163; females, 163–64; fertilization, 163; males, 163–64; nests, 163, *163,* 285
Wrens, 201–04; cactus, 204; use of camouflage, 202; females, 202, 210; males, 202, 210; nests, as roosts, 202–04; roofed nests of, *194,* 202–04, *203,* 218; winter, 201

Index of Scientific Terms

Acanthaster, 12
Acrocephalus, *176*, 197
Acropyga, 121
Actinia, 10
Actinomma, 8
Agapornis, 189
Agroeca, 34
Alcedinidae, 224
Alcedo, 223–24
Alectura, 177
Alligator, 170
Amblyornis, 242
Amitermes, 138
Ammophila, 52, 55
Amoeba, *3*, 4
Anabolia, 48
Anas, 185
Anthoscopus, 206
Anthozoa, 10
Apicotermes, 143, 144
Apidae, 73, 100
Apis, 95
Apodidae, 228
Aptenodytes, 231
Apterona, *47, 48*
Apus, 228
Aquila, 197
Arachnida, 24
Arachnocorys, *8*
Ardea, 192
Argyroneta, 34
Arthropoda, 24
Assilina, *6*
Astraea, *20*
Atta, *102*, 104, 116–17, *118*
Auriparus, 206
Australopithecus, 22

Basicladus, 48
Bembex, 51, 55
Betta, 159
Bombus, 73, 75–76, *79*

Botaurus, 192
Bubalornis, 211
Bucerotidae, 221
Bucorvus, *222, 223*

Cacicus, 217
Camponotus, *99*, 110, 114, 115
Campylorhynchus, 204
Carchesium, *20*
Cardium, 160
Castor, 266
Chalicodoma, 69–70, *70*
Charadrius, 184
Charonia, 20
Chartergus, 64, *77*
Chiromantis, 165
Chlamydera, 239, 240
Chrysopa, *40*
Cisticola, 227
Cladosporium, 110
Colaptes, 221
Colibri, 190
Colletes, 67
Collocalia, 228
Colobopsis, 103, 110
Coptotermes, *126*
Corallium, 13, *18*
Cornitermes, 137, *138*
Crematogaster, 120, 217
Crenilabrus, 162, *163*
Crocodylus, 170
Crustacea, 24
Cryptotermes, 128
Ctenizidae, 35
Cubitermes, 138, *139*
Cyclophorus, 20
Cygnus, 187
Cypsiurus, 230

Dendrocopus, *233*
Dermochelys, *154*

Difflugia, 4
Dryocopus, 189, 220

Eciton, 122
Epibembex, *40, 51*
Eucyrtidium, 8
Eumenes, 55
Euplectella, 9, *17*
Euplectes, *215*
Eutermes, 146
Evylaeus, 72

Foraminifera, 4–6, 7, 9
Formica, 96, *98, 102*, 107, 108, 109, 122
Formicidae, 101
Fungia, 10
Furnariidae, 225

Gasterosteus, 161
Gastropoda, 15
Gerygone, 217, *218*
Gliridae, 258
Globigerina, 7
Gnathypops, 164
Gobiidae, 159
Gobius, 159, *160*
Grammotaulius, *48*
Gygis, 183

Haliaëtus, 197
Halictus, 68, 72
Helicopsyche, 43–44, *48*
Helix pomatia, *14*
Heriades, 67
Hodotermes, *126*, 128
Hydra, 9
Hyla, 167
Hylaeus, 67
Hylocharis, 190, 199
Hymenoptera, 51, 66, 101, 124, 125, 150

Icteridae, 217
Insecta, 24
Isoptera, 124
Ixobrychus, 192

Kalotermes, 128
Kitagamia, 48

Labridae, 162
Lambis, 20
Lasius, 96, 109, 110, 120
Legatus, 191
Leipoa, *155*, 179
Limnophilus, 48
Lithoptera, 8

Macropodus, 157
Macrotermes, 125, 132, *133,*
    135, 139, 142, 143
Malanerpes, 221
Malimbus, 207, *214*
Marmota, 263, 265–66
Megachile, 68
Megalomma, 16, *17*
Megapodiidae, 172
Megapodius, 178
Meles, 251
Melipona, 96
Meliponinae, 96
Messor, 115
Metapolybia, 65
Metisa, 48
Microlepidoptera, 45–47, *48*
Micromys, 254
Miliola, 7
Murex, *16*, *20*
Muscardinus, 258
Muscicapidae, 227
Mustelidae, 251
Myrmecia, 106
Myrmecocystus, 116
Myrmeleon, *24*, *25*

Nasutitermes, *126, 145*
Nemesia, 35, *38*
Neotoma, 258–59
Nephila, 27, 28, *37*
Neureclipsis, *42*
Neuroptera, 25, 50
Nummulites, *7*

Odontotermes, 139, 142, 146
Odynerus, 56
Oecophylla, 111, *112*, *113*,
    115
Oiketicus, 48
Opisthognathidae, 164
Oplomerus, 56
Orthotomus, 227
Osmia, 70–71, *71*

Pandion, *174*, 197
Panyptila, 230
Paridae, 204
Passer, 203
Passerinae, 207
Peneroplis, *7*
Pheidole, *102*
Philetairus, 211, *212*
Phryganea, *44*
Phylloscopus, *194*, 202
Picidae, 218
Plethodontidae, 165*n*.
Ploceidae, 207
Ploceinae, 207
Ploceus, 211
Podiceps, *174*, 192
Pogonomyrmex, 115
Polistes, *60*, 65
Polybia, 63–64, 77
Polychaeta, 16
Polyrhachis, 114, 115
Polystomella, *5*
Pseudonigrita, *213*
Psychidae, 47–48, *48*

Ptilonorhynchidae, 237
Ptilonorhynchus, 238

Radiolaria, 6–7, 9
Rana, 169
Remiz, 204
Remizidae, 204, 206
Reticulitermes, 128, 146–47
Rhacophorus, 165
Rhinoderma, 168

Scala, 20
Sciurus, 259, 262
Sericostoma, *45*
Sitta, 191
Somateria, 185
Sphecidae, 51
Spheniscidae, 231
Struthio, 184
Sylvia, *193*, 197, 198
Sylviidae, 197

Talpa, 248
Tamandua, 131, *136*
Tamiasciurus, 262
Taxidea, 253
Termitidae, 127
Termitomyces, 137
Textularia, 7
Tibia, 20
Triaenodes, 48
Trichoptera, 41, *48*
Trinervitermes, 143, 146
Trochilidae, 199
Troglodytes, 201
Troglotydidae, 203
Tyrannidae, 191

Uria, 183

Vespidae, 55